Vaughan Cornish

Practical Proofs of Chemical Laws

A course of experiments upon the combining proportions of the chemical

elements

Vaughan Cornish

Practical Proofs of Chemical Laws
A course of experiments upon the combining proportions of the chemical elements

ISBN/EAN: 9783337275815

Printed in Europe, USA, Canada, Australia, Japan

Cover: Foto ©berggeist007 / pixelio.de

More available books at **www.hansebooks.com**

PRACTICAL PROOFS

OF

CHEMICAL LAWS

A COURSE OF EXPERIMENTS
UPON THE COMBINING PROPORTIONS OF
THE CHEMICAL ELEMENTS

BY

VAUGHAN CORNISH, M.Sc.

ASSOCIATE OF THE OWENS COLLEGE, MANCHESTER

LONDON
LONGMANS, GREEN, AND CO.
AND NEW YORK
1895

PREFACE

THESE experimental proofs (or more properly *verifications*) of quantitative laws were undertaken by pupils after the qualitative composition of the principal substances employed had been carefully dealt with in the accompanying lecture course.

Practical Physics went on side by side with the practical chemistry course.

The pupils whose results are quoted in the text were mostly between twelve and eighteen years of age. A book of results was kept so that each pupil could compare his results with others obtained under similar conditions.

The pupils had $1\frac{1}{2}$ hour at a time in the laboratory, and attended twice a week.

I have not been satisfied with quantitative experiments unless they yield good results in the hands, not only of the teacher but of the pupils. The results quoted in the text are those obtained by the pupils.

I am not aware that a satisfactory standard as to the accuracy required for such experiments has yet been laid down. Within 1 per cent. is certainly sufficient, but the standard may vary to some extent according to the nature of the law or problem investigated.

Perhaps the standard is best determined by historical considerations, for the history of a science is recapitulated by the learner. If the pupil can verify a law to such a degree of approximation as first served to convince the scientific world of its truth, he may generally be satisfied with his work. I have quoted in the text the results of early historic experiments side by side with those obtained by pupils. Although in these early experiments the error is often large, yet there is less difference than the learner may have supposed between the accuracy of the *first approximations* which have obtained the provisional assent of the scientific world at the beginning and towards the end of the nineteenth century, respectively.

As far as possible no numerical data, whether chemical or physical, were *assumed*. The density of hydrogen and the proportion by weight in which hydrogen and oxygen combine are neither assumed nor determined in these experimental verifications of the Laws of Combining Proportions. It is not *necessary* that equivalent weights should be referred to that of hydrogen, and the experiments cannot be done with the same accuracy as is attainable in the case of other elements.

The use of atomic and molecular formulæ is inadmissible in an examination of the facts upon which the atomic and molecular theory is based. Chemical equations and formulæ have therefore been excluded.

The course is, I believe, suitable for first-year's students at colleges as well as for the upper forms of schools.

I have to acknowledge valuable suggestions received from other chemists while this course was in preparation. My thanks are due more particularly to Dr. J. B. COHEN, of the Yorkshire College, Leeds, and to Mr. G. STALLARD, of Rugby.

VAUGHAN CORNISH.

August 1895.

CONTENTS

———◦◇◦———

CHAPTER I

CHAPTER II

THE LAW OF CONSERVATION OF MASS

CHAPTER III

THE FIRST PART OF THE LAW OF DEFINITE
AND OF CONSTANT PROPORTIONS

CHAPTER IV

THE LAW OF EQUIVALENT PROPORTIONS

CHAPTER V

THE SECOND PART OF THE LAW OF DEFINITE
AND OF CONSTANT PROPORTIONS

Contents

CHAPTER VI

THE LAW OF MULTIPLE PROPORTIONS

CHAPTER VII

THE LAW OF SIMPLE VOLUMETRIC PROPORTIONS IN THE CHEMICAL REACTION OF GASES

Contents

CHAPTER VIII

PRACTICAL

.

Errata.

Page 10 line 23 *for* 2 to 3 cm.   *read* ·2 to ·3 cm.
,, 35 ,, 28 ,, 50 cb.c.   ,, 500 cb.c.
,, 42 ,, 30 ,, 55 p. c;   ,, ·55 p. c.
,, 42 ,, 32 ,, 60 p. c.   ,, ·60 p. c.
,, 73 ,, 28 ,, hypobromide ,, hypobromite.

ωeory of chemical action may
ᴗd as follows :—
When a chemical action takes place, what we observe on the large scale is the total effect of a vast number of similar actions occurring between ultimate particles, or atoms, of the substances. The atom of each chemical element has its own specific mass. These chemical atoms are beyond our powers of vision, and we have no means of dealing with them individually. Dalton's theory remains therefore a theory only, and has not been raised to the rank of a statement of observed facts. The theory is, however, based upon observed facts ascertained by experiment.

B

CHAPTER VIII

EXERCISES SUPPLEMENTARY TO THE COURSE ILLUS-
TRATING THE SCOPE OF THE TERM EQUIVALENCE
IN CHEMISTRY

PRACTICAL

PROOFS OF CHEMICAL LAWS

——◆◇◆——

CHAPTER I

STATEMENT OF DALTON'S ATOMIC THEORY AND
OF THE LAWS OF COMBINING PROPORTIONS

DALTON'S atomic theory of chemical action may
be stated as follows :—

When a chemical action takes place, what we
observe on the large scale is the total effect of a
vast number of similar actions occurring between
ultimate particles, or atoms, of the substances. The
atom of each chemical element has its own specific
mass. These chemical atoms are beyond our
powers of vision, and we have no means of dealing
with them individually. Dalton's theory remains
therefore a theory only, and has not been raised to
the rank of a statement of observed facts. The
theory is, however, based upon observed facts
ascertained by experiment.

B

The laws of chemical combination by weight are the experimental basis of Dalton's theory.

The later discovery, by Gay-Lussac, of the simple ratios between the reacting volumes of gases (the 'Laws of Combination by Volume') led to the development of the atomic theory in its present form, in which we suppose the existence of two orders of particles, the *molecule* and the *chemical atom.* The modern development of theoretical chemistry is due in great measure to the theory of atoms and molecules, the theory itself being based, as has been said, upon the laws of combining proportions, gravimetric and volumetric. These are both included in the term *Laws of combining proportions.*

The methods of chemical analysis are founded directly upon these experimental laws and upon the law of conservation of mass, the results obtained by analysis being independent of the atomic theory.

These laws are therefore the foundation of the greater part both of practical and of theoretical chemistry.

We proceed to enunciate the laws before giving the description of experiments by which each of them may be verified.

1. THE LAW OF CONSERVATION OF MASS.— The law, in its bearing upon chemistry, may be stated in a general form as follows :—

'The total mass of the substances taking part in any chemical process remains constant' (*Ostwald*).

We will for the purpose of this course state the law in a less general form, in which it can be readily verified, as follows :—

When elements combine together chemically, the mass of the compound formed is equal to the sum of the masses of the elements before combination.— As *mass* is almost always measured by *weighing*, we shall substitute the more familiar term *weight* in the following statements of laws.

2. THE LAW OF DEFINITE AND OF CONSTANT PROPORTIONS. PART I. (*definite proportions*).-- *When two elements combine to form a particular compound substance, they do so in a definite, fixed proportion by weight, which is independent of the manner in which their combination is brought about ; and* PART II. (*constant proportions*). *This proportion remains constant in compounds which contain also other elements.*

3. THE LAW OF EQUIVALENT PROPORTIONS. *The weights of two elements which are equivalent* (i.e. *of equal value*) *in any chemical reaction are equivalent in all.*

The meaning of this general statement of the law may be illustrated by special statements applicable to particular cases which require different experimental methods for their verification.

First special statement (applicable to the case of elements each of which is capable of combining with each of the others, e.g. silver, chlorine, and sulphur):

The weights of two elements (e.g. sulphur and

chlorine) *which combine with a certain fixed weight of a third element* (e.g. silver) *are in the proportion in which those two elements* (sulphur and chlorine) *combine with one another.*—These weights of sulphur and chlorine are said to be *equivalent* to one another.

Second special statement (applicable to the case of elements, some of which do not combine together, e.g. silver, potassium, chlorine and bromine) :

The weights of two elements (e.g. silver and potassium) *which combine with a given weight of a third element* (e.g. chlorine) *will also combine with another fixed weight of a fourth element* (e.g. bromine).—The experimental verification of this statement of the law is given in Chapter IV.

4. THE LAW OF MULTIPLE PROPORTIONS.— It sometimes happens that there are two (or more) different substances formed from the same elements. By different substances we mean materials which differ in a marked degree in their physical characters (e.g. in density, boiling point, melting point, and so forth). To such cases the law of multiple proportions applies. To simplify the wording of the law, we will frame a statement suited to the case of substances containing only two elements.

Statement of the Law of Multiple Proportions.— If there be more than one substance formed by the combination of two elements, then, taking the weight of one element as fixed in each substance, the weight of the other element in the second compound bears a simple proportion to the weight of that element in the first.—When the student has fully grasped the meaning of the word equivalence in chemistry, which is generally not until he has had some experience of experimental work, he will find the following statement useful as a summary of all the laws of chemical combination by weight:

Elements combine together in the proportion (or ratio) of their equivalent weights, or in the proportion of whole multiples of their equivalent weights.

5. THE VOLUMETRIC LAW OF COMBINATION OF GASES may be stated thus:—*The volume of an element in the gaseous state bears a simple proportion to the volume of the compound gas of which it is a constituent.*

CHAPTER II

THE LAW OF CONSERVATION OF MASS

STATEMENT.—*When elements combine together chemically the mass of the compounds formed is equal to the sum of the masses of the elements before combination.*

MASS is the one property of matter which remains absolutely constant in every state of chemical combination. The Balance is the instrument employed in the comparison of masses, and a knowledge of the systematic method of weighing must be acquired before undertaking quantitative chemical experiments.

EXERCISE I.—*The Use of the Balance.*

Apparatus required.—A chemical balance, the pointer of which will move one division on the scale when 1 mgrm. is placed on the empty pan, and which will bear a load of 50 grams on each pan. Becker's 50s. balance fulfils these conditions. A set of weights from 50 grams to 1 mgrm. It is well that each pupil should have a set of fractions of a centigram. These can be obtained, in platinum, for about 2s.

Determination of the Zero point.—Raise the
beam of the balance from its support by turning
the knob which is on the front of the balance
case. The beam now swings freely. The balance
is adjusted by the instructor, not by the pupil, so that
the pointer swings *nearly* evenly on either side of
the central line on the ivory scale. It is required
to determine exactly the point on the scale which
marks the position of equilibrium, or the *Zero
point* of the balance. This is best done, not by
waiting for the pointer to come to rest, but by
observing the swings on either side of the central
point. The amplitude of the swings steadily
diminishes owing to friction. Thus taking 10 as
the value of each division, calling the central point
100, and reckoning through from left to right, we
may observe the following extreme positions of
the pointer :

Left hand	Right hand
85	
	110
87	

the parts of a division being estimated by eye.
Now 85 is further from the Zero point and 87 is
nearer to the Zero point than 110. We may
assume that $\frac{85 + 87}{2} = 86$ is as far on the left of
the Zero point as 110 is on the right. The Zero
point is found by taking the mean of the two
numbers 86 and 110 ; thus :

$$\frac{86 + 110}{2} = 98 \text{ (Zero point)}$$

Determination of Sensibility.—In weighing a body it will generally be found after adjusting the weights as nearly as possible that the pointer swings nearly, but not quite, evenly about the Zero point. The weights being placed on the right-hand pan, and the body to be weighed on the left-hand pan, then if the Zero-point is 98 and the pointer is found to swing evenly about the position 90, it is evident that the weights are somewhat too heavy. We require to know what extra weight in one pan produces a deflexion of 8, i.e. $\frac{8}{10}$ths of a division. *To determine the sensibility with empty pans*, place one milligram on the right-hand pan and observe three or five swings. Suppose these to be

Right hand	Left hand
126	
	52
122	

the balancing-point is evidently

$$\frac{124 + 52}{2} = 88$$

The Zero point being 98, the sensibility with a very small load is 10. Thus, in the example given, if the weight of the body was small (say 5 grams) we conclude that the weights on right-hand pan were too heavy by $\frac{8}{10}$ths of a milligram. The true weight of the body is therefore obtained by deducting ·0008 gram from the amount of the weights on the right-hand pan.

To determine the sensibility with a load of 20 *grams in each pan.*—The sensibility generally decreases with heavier loads. In the experiments given in this course the loads are generally either small (less than 5 grams) or about 20 to 25 grams in each pan—e.g. in weighing porcelain crucibles. To determine the sensibility in this second case, place a 20-gram weight in each pan and determine the balancing-point. This may not be exactly the Zero point, since the two weights may not be exactly equal. Suppose the balancing-point to be 96. Place ·001 gram (1 milligram) on the right-hand pan and again determine the balancing-point. Suppose this to be 88. The sensibility with a load of 20 grams in each pan is 8.

Record of the results of the exercise.—The observations of swings and the results for sensibility are to be clearly stated in the pupil's note-book. They should be taken down in the first place in a small pocket note-book, in which all weighings, as well as observations made during the actual progress of experiments, should be entered at the time, to be copied out afterwards in a larger note-book. Weighings should on no account be entered on loose sheets of paper ; the results of experiments are often lost in this manner.

EXERCISE II.—*Verification of the Law of Conservation of Mass by the complete synthesis of silver sulphide.*

This experiment shows that when silver and sulphur are heated together, they combine chemically, producing a new substance (silver sulphide), and that *the weight of the silver sulphide formed is equal to the sum of the weights of the silver and sulphur from which it was produced.*

Apparatus and substances required.—*Fine silver wire* on a reel—that used in surgery is the best. *Re-sublimed sulphur*, as obtained from the dealers, but freed from moisture by being kept on a clock glass in a desiccator over strong sulphuric acid. The clock-glass must be replaced in a desiccator when a portion of the sulphur has been taken out for the experiment. *Glass tubing* of soft glass, not combustion tubing ; the most convenient width of bore is 1 to 1·5 cm. It should be sealed up at one end and be of a length not less than 16 to 20 cm. *A cork* to fit this tube, through which passes, somewhat loosely, a piece of glass tubing about 2 to 3 cm. diameter. *Carbonic acid apparatus. Foot-blowpipe and bellows.* An *oven or gas-furnace* in which a tube of about 10 cm. length can be placed and heated to about 500° C. In default of this, the tube may be covered with sand and heated on a sand-tray. *Weighing-tube, watch-glass, glazed paper, camel-hair brush* or, better, a *stiff feather, carbon bi-sulphide, sharp three-cornered file, scissors.*

Method of conducting the experiment.—Make the glass tube thoroughly dry by moving it from side to side and rotating in a luminous gas flame, at the same time blowing air through from the bellows. Measure off a length of the silver wire which has been found to weigh about one gram. Cut this up with scissors into short pieces, about ·5 c.m. length, and weigh accurately. This substance may be weighed on the bare pan of the balance. Transfer the silver wire to the tube. We now have to weigh out *about* enough sulphur to combine with the whole of the silver. For 1 gram silver take ·15 to ·20 grams sulphur. The quantity taken must be accurately weighed. The sulphur may be placed in a small stoppered weighing tube which must be dry. This is weighed with its contents, and the sulphur is then shaken out carefully, little by little, into the tube which contains the silver. Should any sulphur fall outside the tube it is received on the glazed paper and can be brushed into the tube. If the tube has not been properly dried the sulphur will adhere to the sides. The weighing-tube which contained the sulphur is corked and weighed again, the difference of the two weighings being the weight of sulphur taken.

If the tube were sealed up while containing air some of the sulphur would burn on heating. In order to avoid this complication the air is displaced by carbonic acid before sealing. The gas is passed in through the narrow glass tube carrying the cork, the cork of course not being fitted into the wider tube. The current of gas is turned on slowly in order to avoid the risk of scattering the sulphur over the sides of the tube. The tube is full of carbonic acid when a taper is extinguished at its mouth. Now stop the current of carbonic acid and fit in the cork ; draw the narrow tube through the cork so that it projects not more than 1 to 2 c.m. into the wider tube. The wider tube must now be sealed up. In the operation of sealing, begin by warming the tube above the luminous flame of the blow-pipe ; then bring it into the luminous flame, keeping it constantly turned. When the tube is covered with soot the blast may be put on, but very gently at first. When the tube softens draw it out slightly, then lower the flame and heat strongly, rotating the tube during the operation, so that the sides fall together. When this has taken place cool slowly in the luminous flame until the tube is again coated with soot. If the pupil has no previous practice in glass-working this part of the exercise should be *rehearsed* with an empty tube to master the manipulation. This practice of rehearsing may with advantage be adopted whenever a quantitative experiment involves a new piece of manipulation. Everything is now ready for bringing about chemical combination by heating. The method of doing this has already been described. The time for completing the reaction depends partly upon the thickness of the wire. It is best to arrange matters so that the weighings and sealing are completed at the end of

a lesson, so that the heating of the tube may be
left to go on for a few hours in order to ensure
completion.   The remaining determinations can be
done in the next lesson.

After heating, it is found that the silver and
sulphur have been converted into the black, shining
crystalline silver sulphide.   If an excess of sulphur
was taken this will be found collected on the part
of the tube which was coolest.   We now have to
weigh the silver sulphide and the excess, if any, of
sulphur.   With a sharp file make a deep transverse
cut at the middle of the tube.   Touch this with a
fine point of glass which has been heated till quite
soft in the blow-pipe flame.   In this way the tube
may be readily cut into halves.   The sulphide is
readily detached from the tube and is brought on
to a weighed watch-glass and weighed.   The tube
must be carefully brushed out, the watch-glass being
placed on the sheet of glazed paper.   The accuracy
of the results obtained in this experiment depends
upon the attention given to the details of manipulation, the chemical part of the experiment not
presenting any difficulty.   If there is any excess
of sulphur, weigh the pieces of glass to which the
sulphur adheres, dissolve off the sulphur with
carbon bi-sulphide, in a draught chamber, and
weigh again.   The difference gives the weight of
the sulphur.

Calculation and statement of results.—Examples
are given from the results obtained by two pupils.
The weights are in grams.

The first pupil found

Before heating		After heating	
Silver	. . ·8301		
Sulphur .	. ·1233	(no excess of sulphur)	
Sum	. . ·9534	Sulphide	. ·9531

The weight of the sulphide is found to be less than that of the elements from which it was formed by 3 parts in 9534. Regarding the exercise as intended to verify a law believed to be mathematically exact, we may say that this represents the *experimental error*. This should always be expressed as a percentage. To find the percentage in the above case divide 3 by 9534 and multiply by 100. The result is, very nearly, ·03. It must be written with the *minus* sign, since the weight of the compound formed is found to be less than it should be ; therefore we write

Experimental error − ·03 per cent.

The second pupil found

Before heating		After hea ing	
Silver	. . . 1·0000		
Sulphur taken .	. ·2963	Excess sulphur .	·1476
Excess sulphur left .	·1476	Sulphide .	. 1·1476
Therefore sulphur used in combination	. ·1487		

Sum of weights of elements before combination	. 1·1487
Weight of compound formed	. 1·1476
Difference	. ·0011
Therefore experimental error	. − ·10 p. c.

The average error of thirteen experiments recorded by ten pupils was ·5 per cent. One of the experiments exceeded 2 per cent. in error, and another exceeded 1 per cent. In no case was the error $+^{tive}$, i.e. the weight of the compound was never found to be *greater* than the sum of the weights of its constituents.

Note upon early experiments relating to the Law of Conservation of Mass, and upon Stas' determinations.—W. Black (' Experiments upon Magnesia Alba,' published 1782, pp. 66-68) converted 120 grains of chalk, by heating, into quicklime, of which 68 grains were left. The lime was thrown into a solution of carbonate of soda, by which means chalk was again formed, which weighed 118 grains. The final weight was, therefore, found to be less than the original weight by 2 parts in 120 : experimental error — 1·6 per cent.

Lavoisier (' Elements of Chemistry,' translated by Dr. Kerr, published 1792) decomposed mercuric oxide by heat, weighed the mercury obtained, and measured the volume of oxygen. He also determined the density of oxygen. The experimental error appears to have been about — 1 per cent., but the statement of results leaves some uncertainty. Lavoisier stated (*loc. cit.*) the law of conservation of mass in connection with his experiments upon the fermentation of sugar in presence of water and yeast. He found that the weight of the products formed was equal to the weight of the substances decomposed. The numbers given are apparently not those actually determined, but rather what he considered they ought to have been

if experimental error were eliminated.   The numbers, therefore, are not available for calculating the
approach to accuracy attained in the experiment.

For modern work upon this law, see Stas'
' Nouvelles Recherches sur les Lois des Proportions
Chimiques,' experiments upon the complete synthesis of silver iodide, and complete analysis of
silver iodate, which confirmed the law within a very
small margin of experimental error.

CHAPTER III

THE LAW OF DEFINITE PROPORTIONS

STATEMENT OF THE LAW.—*When two elements combine to form a particular compound substance, they do so in a definite, fixed proportion by weight, which is independent of the manner in which their combination is brought about.*

THIS is the *first part* of the law of definite and of constant proportions ; for the *second part*, see Chapter V.

The significance of this law lies in the *persistence* with which elements adhere to certain proportions of combination in spite of variation in the conditions under which combination occurs. The conditions of combination may vary in the following respects :—

1. Relative masses of the reacting substances.
2. Pressure under which the reaction occurs.
3. Temperature at which the reaction occurs.
4. Chemical composition of the substances by the reaction of which the combination is brought about.

The fact that the combining proportion does not depend upon the relative masses present may

C

be verified by a comparison of the results obtained
by different pupils in the last exercise. Thus,
taking the two examples already quoted :—

	Relative masses taken	Combining proportion
First pupil .	Sulphur 1 : Silver 6·732	Sulphur 1 Silver 6·732
Second pupil .	,, 1 : ,, 3·333	,, 1 ,, 6·777

the difference between the combining proportions
obtained by the two pupils is ·66 per cent., the
combining proportions being constant within ·66
per cent. when the proportion between the substances taken was varied from 2 to 1.

The second condition (variation in pressure) is
tested to some extent by the same experiment,
since the pressure in the tube depends, other things
being equal, on the excess of sulphur present.

The third and fourth variations of condition
(viz. of temperature, and of the chemical composition of the reacting substances) are tested by the
following experiments upon the proportion by
weight in which silver and chlorine combine
together. The last (fourth) mode of varying the
conditions is especially important as showing that
the combining proportions are independent not
only of physical but also of chemical conditions.

In the first experiment (Exercise III.), the elementary gas chlorine acts on metallic silver at a
temperature of about 400° C. to 500° C. In the
second experiment (Exercise IV.) the combination
of the two elements, chlorine and silver, is brought
about by the action of a solution of hydrochloric
acid upon silver nitrate at a temperature rather

below 100° C. Exercises III. and IV. may be worked side by side. Such an arrangement as the following will save time. Weigh out the materials and start the evaporation in Exercise IV. While this is going on, weigh out for Exercise III. and start the heating in chlorine. Then return to Exercise IV., which can be finished. The chlorination in Exercise III. is, if possible, left going for some hours, the weighing and calculation being postponed till the next lesson.

EXERCISE III.—*Determination of the proportion in which silver and chlorine combine when chlorine gas acts upon metallic silver.*

The following are the *apparatus and substances required, and the method of conducting the experiment.* Precipitated *silver* is used, not less than ·5 gram. The *heating arrangement* to be adopted will depend partly upon the equipment of the laboratory. Two sources of error must be provided against :—*First*, portions of silver may become surrounded by fused chloride, and thus be kept from the action of the gas. To avoid this as much as possible, the metal should present a large surface, the stream of gas should be slow, and the temperature should be regulated so that the chloride only just fuses. *Secondly*, chlorine gas is absorbed by the fused chloride, and on cooling the escape of the gas is apt to cause spirting. This is to some extent guarded against by the precautions already men-

C 2

tioned, and also by cooling slowly at the end of
the operation. In order to ensure that no loss
should take place even if spirting did occur, either
of the following arrangements will do. *First
arrangement of apparatus :*—The silver and the
resulting chloride are weighed in a glass tube,
closed at one end, provided with a loose plug of
glass wool about half-way up, and fitted with a
two-way stopper, the chlorine entering through one
tube, and the excess of the gas passing out of the
other tube to the draught. The closed end of the
tube where the charge of silver lies can be heated
by means of a Bunsen burner. This arrangement
can only be adopted conveniently if the balance is
capable of taking a tube of the necessary length.
Second arrangement of apparatus :—If a short combustion furnace be available, the silver may be
weighed in a porcelain boat, which is placed inside
a glass tube lying in the bed of the furnace. If
any spirting take place the particles of chloride
may be detached from the glass tube and returned
to the boat after the operation is at an end. The
exit end of the tube is connected to the draught.

The chlorine apparatus should supply a regular
stream of the gas for some hours without requiring
attention. It should not require replenishing
during use ; in fact, it should be of such a form
that it needs replenishing only at long intervals.
The Kipp apparatus, with balls of compressed
bleaching powder, is readily set working, and has
the advantage of not requiring heat. It has, how-

ever, the disadvantage that when left standing the
joints are subjected to pressure, and any leak
results in the liquid running down. It also needs
replenishing more often than the *porcelain chlorine
still.* The chlorine still, made by the Berlin
Porcelain Company, is filled to a depth of about
4 c.m. with pebbles or glass stoppers, and on these
are placed lumps of manganese dioxide, 1–1·5 cb.c.
in diameter. The hydrochloric acid used is of the
strength of equal parts of the strong acid and of
water. The joint between the cover and the vessel
is made tight by a number of filter papers soaked
in oil. The porcelain vessel stands in a water-
bath, which is supplied with it, and when a current
of gas is required a burner is lighted below the
water-bath. The water is soon warmed, as the
surface of the metal is large and the volume of
water is relatively small. It is best not to allow
the water to boil, as the steam is apt to cause the
filter-paper joint to leak. A glass stop-cock con-
trols the flow of gas from the still, and a screw-tap
or nipper-tap should be used to control the entrance
of the acid from the upper vessel. The glass stop-
cock having been partly turned on, matters are
adjusted before leaving the apparatus to itself, so
that the level of the acid in the upper vessel remains stationary. When it is desired to stop the
current the exit tap is closed. The amount of
liquid in the still can be judged by the level of the
acid in the upper vessel. If most of the acid is in
the upper vessel, turn out the burner under the

water-bath and close the inlet tap.   If, on the other
hand, most of the acid has run down into the still,
continue heating after turning off the exit tap until
most of the acid is driven into the upper vessel,
when the inlet tap may be closed.   When the acid
requires renewal, syphon from the upper vessel,
the inlet-tap being closed.   A large supply of
manganese dioxide can be put into the still,
sufficient to last without renewal for a term's
laboratory work.   The still can then be replenished
when the laboratory is no longer in use by the
pupils, and the apparatus can be got ready again
(air expelled, acid driven back, and joints tested)
immediately before the commencement of the next
term's laboratory classes.

For calculation and statement of results of the
exercise see *post*, at the end of Exercise IV.

EXERCISE IV.—*Determination of the proportion in
which silver and chlorine combine when a solution
of hydrochloric acid acts upon a solution of silver
nitrate.*

The following are the *apparatus and substances
required :*—Granulated *silver*. *Nitric acid,* the 'pure,
strong' should be used, but a little hydrochloric acid
as impurity does not matter. *Hydrochloric acid,*
pure, strong. *Porcelain crucible and lid* of the size
generally used in quantitative analysis, weighing
about 22 grams. *Pipette* delivering 10 cb.c. *Iron
tripod, pipe-clay triangle, crucible tongs,* preferably

of gun-metal, *desiccator*, *wash-bottle*, and *water-bath*. Water-baths are somewhat expensive, and economy may be effected in working this course of experiments, in which the evaporations are done in crucibles, by using water-baths such as those made for Wanklyn's process of milk analysis, which are provided with a number of holes of a size required for crucibles. As it is desirable that the evaporation should not require watching, the bath should be furnished with an arrangement for keeping the level of the water constant.

Method of conducting the experiment.—Weigh a porcelain crucible and lid. Weigh out in the crucible ·5–·6 grams pure granulated silver. Just cover with hot water and add strong nitric acid from a pipette *slowly*, so as to keep up a fairly rapid evolution of gas without such effervescence as might result in loss. Towards the end the process may be hastened by warming on the water-bath, but the contents of the crucible must not be evaporated to dryness while any silver remains undissolved. When the whole of the silver has dissolved, evaporate to dryness in order to expel excess of nitric acid, the crucible being uncovered and placed in a draught chamber during the process. Dissolve the silver nitrate in a very little hot water, add 6–10 cb.c. strong pure hydrochloric acid, and evaporate to complete dryness on the water-bath. When dry, replace the lid and heat gently on the pipe-clay triangle, preferably with a rose on the Bunsen burner. When there is no

longer danger of spirting, remove the lid and heat
the crucible carefully with a small flame of the
Bunsen burner, applying the heat round the sides
of the crucible rather than at the bottom.   As soon
as the chloride begins to fuse round the edges,
cease heating, replace the lid, and transfer to the
desiccator.   Weigh when quite cold.

Calculation and Statement of Results of Exercises III. and IV.

A pupil found that

·500 grams silver by second method gave   ·   ·664 of chloride

and that

·497 grams silver by first method gave   .   ·659 of chloride ;

therefore

·500 grams silver by first method would give   ·663 of chloride.

Therefore, assuming the truth of the law of conservation of mass, which was verified in Exercise
II., the proportions in which chlorine combines
with a fixed weight of silver under the different
conditions of Exercises III. and IV. were found to
be as

$$163 : 164$$

The difference is ·6 per cent., which, on the assumption that the law of definite proportions holds
exactly, is the experimental error.

For a supplementary experiment (Exercise
IV. A.), illustrating the application of the rule of

definite proportions to the case of elements which do not combine together, see end of this chapter.

For an experiment verifying the *second part* of the law (viz. that the definite fixed proportion between the weights of two elements combining to form a particular compound substance is a *constant* proportion between the weights of those two elements present in chemical compounds containing also other elements) see Chapter V.

Note upon early experiments relating to the Law of Definite Proportions, and upon Stas' determinations.—The accumulation of the results of chemical analysis in the later part of the eighteenth century gradually established the fact that a number of well-known substances, e.g. certain salts and minerals, had a definite fixed composition. Some chemists were disposed to regard the fixed composition of these substances as evidence of a general law that elements can only combine chemically in certain constant proportions which are independent of the conditions of combination. The growth of this view was strongly combated by Berthollet in the first years of the nineteenth century. He contended that there was no such restriction upon the combining proportions of the elements. The fixity of composition of characteristic chemical substances he considered to be determined mainly by conditions, such as insolubility, volatility, &c. Analyses were undertaken by Proust, designed to test the view that the combining proportions of chemical elements are independent of conditions of

temperature, pressure, solubility, and so forth.
The issue of the controversy confirmed the views
advocated by Proust in opposition to Berthollet.
We quote some of Proust's results. He writes
(' Journal de Physique,' vol. lv., A.D. 1802, p. 326) :
' One hundred parts of antimony, and the same
amount of sulphur, heated in a glass retort till the
whole is completely fused and the excess of sulphur
has been driven off, leave 135 parts by weight of
the sulphide.

' This experiment, however often repeated,
always yields the same result.   One hundred parts
of antimony heated with 300 parts of cinnabar (the
native sulphide of mercury) give from 135 to 136
parts of the sulphide.   These sulphides heated with
an equal weight of sulphur did not increase in
weight.   It follows, that antimony conforms to the
same law as all the metals which are capable of
combining with sulphur.   They take up a constant
quantity fixed by Nature, and Man has no power
to increase or diminish this quantity.'   This result
may be taken as verifying the Law of Definite Proportions within about ·5 per cent.

In dealing with the more difficult case of iron
pyrites (' Journal de Physique,' vols. lii. and liv.),
Proust's results are less satisfactory from a
numerical point of view, although they are more
interesting in so far as they afford an early example
of the artificial reproduction of a natural mineral,
and a *rough* confirmation of the fact that the amount
of sulphur with which iron can combine is the same,

whether the sulphide be formed in the laboratory or in, e.g. a mineral vein. Proust found that 400 parts of the natural iron pyrites were .reduced by heating sufficiently to 318 parts. He then mixed 318 parts of the common black sulphide of iron with sulphur, and after heating, not too strongly, found that a substance was formed having the principal properties of the natural iron pyrites and which weighed 378, instead of 400. This confirms the law to within about $5\frac{1}{2}$ per cent.

In 1805 Gay-Lussac and Humboldt found that the proportions in which hydrogen and oxygen combine to form water is unaffected by differences of temperature and pressure.

For modern work showing to a high degree of accuracy that the proportions of combination in the formation of a particular compound are unaffected by circumstances of temperature and pressure, or by the source from which the substances are derived, see Stas' analysis of ammonium chloride ('Nouvelles Recherches sur les Lois des Proportions Chimiques' in the 'Mémoires de l'Académie Royale de Belgique,' vol. xxxv. pp. 48-57).

APPLICATION OF THE LAW OF DEFINITE PROPORTIONS TO THE CASE OF ELEMENTS WHICH REACT, BUT DO NOT UNITE TOGETHER.—To meet the case of the chemical *displacement* of one element by another we may adopt the following statement of the Law of Definite Proportions :—

Chemical elements react together in a definite fixed

proportion.—The displacement or replacement in a definite fixed proportion may be verified by an experiment upon the action of magnesium (or zinc) upon dilute sulphuric and hydrochloric acids, when the metal dissolves in the acid and hydrogen gas is evolved.

EXERCISE IV. (A). *Determination of the quantity of hydrogen evolved on the solution of a given weight of magnesium (or zinc) in dilute sulphuric acid and dilute hydrochloric acid respectively.*

The following are the *apparatus and materials required :*—Either *magnesium ribbon, or*, if the experiment be made with zinc instead of magnesium, *pure granulated zinc* should be used. *Glass tube* about 2 c.m. long and 1 c.m. wide closed at one end. *Glass wool. Pad of caoutchouc* (a slice from a large rubber stopper will do) with a hole cut to fit the closed end of the tube, and a nick or channel in the upper surface. *Glass-mortar* or small pudding-basin, or deep evaporating dish. Pure dilute *sulphuric* and *hydrochloric acids*. Wide *glass cylinder*, or other deep vessel, *thermometer*, and a graduated *gas-measuring tube* which can be closed with the thumb (see also Chapter VII., Exercise XV.). *Retort stand and clamp.*

Method of conducting the experiment.—Weigh out with the greatest care about ·06 gram of magnesium in short pieces (or about ·15 gram granulated zinc). With a balance such as has been referred to in Chapter II., and weighing by vibrations, the error of weighing need not be more than ·5 per cent. of the weight of zinc, but may be as much as 1 per cent. on the weight of magnesium. Place the pieces of metal in the short glass tube. Plug the open end *loosely* with glass-wool. Place the closed end of the tube in the hole cut in the piece of caoutchouc provided for this purpose. Fill up the short tube with water. Half fill the dish or mortar with weak sulphuric (or hydrochloric) acid. The ordinary dilute sulphuric acid used in the laboratory (1 of strong acid to 5 of water) may for this purpose be diluted with an equal bulk of water. Similarly, ordinary dilute hydrochloric acid (1 of strong acid to 3 of water) may be further diluted with an equal bulk of water. Fill the graduated tube with the same acid as that in the basin. Close the open end of the tube with the thumb and invert it in the dish, the dish being placed upon the foot of the retort stand. Hold the graduated tube in a slanting position, and quickly bring the open end of the short tube, which contains the metal, under the open end of the graduated tube. At once bring the graduated tube into the vertical position and clamp it so that it rests firmly upon the pad of caoutchouc. The metal dissolves in the acid with evolution of hydrogen gas. If the plug of glass-wool is too tight the operation is delayed, but it must not be so loose as to be carried away by the rush of gas, otherwise pieces of metal may float to the surface of the liquid and be left adhering to the sides of the measuring tube above the level of the acid. In this case it would be necessary to incline the tube so that the acid flows on to the metal. When the metal is all dissolved transfer the graduated tube to a tall cylinder filled with water. Bring the water to the same temperature, say 15° C., in both experiments. Allow the tube to remain in the water at this temperature for five or ten minutes, and then, holding the tube in a paper holder, or by means of a clamp, to prevent warming by the hand, read off the volume of the gas, the level of the water being made the same inside and outside of the tube. The level of the gas is determined from the level of the bottom of the water meniscus in the tube.

Calculation and statement of results.—Calculate the volume of gas evolved for the same weight of metal, (*a*) when sulphuric acid and (*b*) when hydro-chloric acid is used, and compare the results. Example :—A pupil found that the volumes of hydrogen evolved by the same weight of zinc from hydrochloric acid and from sulphuric acid respectively differed by ·7 per cent.

CHAPTER IV

THE LAW OF EQUIVALENT PROPORTIONS

THE experiments described in this chapter upon the verification of the above law show that *those weights of silver and potassium which combine with a certain weight of chlorine combine also with another fixed weight of bromine.* For a general statement of the law, see Chapter I. We assume, throughout the course, the truth of the Law of Conservation of Mass which was verified in Exercise II.

It is desirable that the pupil, before conducting the experiments described in this Chapter, should be acquainted with the evidence showing that potassium chloride (or bromide) is a compound of potassium and chlorine (or bromine) only. Davy's proof of the composition of potassium chloride may be found in his ' Collected Works,' vol. v. p. 58 *et seq.*, Cavendish Society's Publications.

The first experiment needed for our verification of the law is that of the synthesis of silver chloride, which has already been performed as an exercise on the Law of Definite Proportions. The second is the synthesis of silver bromide, which is shortly described in the next paragraph. The analyses of the chloride and bromide of potassium involves the application of methods not yet explained in this course, and a more detailed description of them is therefore given. If desired, iodine may be substituted for bromine in the following exercises.

EXERCISE V.—*Determination of the combining proportion of silver and bromine, by the action of hydrobromic acid on silver nitrate.*

The following are the *apparatus and substances required:* — Pure granulated *silver*, solution of *hydrobromic* acid (the unsaturated acid containing free bromine will do), pure strong *nitric acid*, which must be free from hydrochloric acid. To test this point, dilute and add a drop of solution of silver nitrate. If a precipitate of silver chloride appears the acid must be purified or a pure sample obtained. To purify, add some silver nitrate solution to the nitric acid and distil from a small retort placed on a sand-tray. The beak of the retort passes into the neck of a glass flask which is kept cooled. The acid which distils over will be free from hydrochloric acid. The *apparatus* required in Exercise V. is the same as in Exercise IV.

Method of conducting the experiment.—Weigh a porcelain crucible and lid. In the crucible weigh out ·25 — ·3 gram of pure granulated silver. Dissolve in the pure nitric acid, evaporate to dryness and re-dissolve in water. Add hydrobromic acid cautiously. If the reaction is violent the lid must be placed upon the crucible to prevent loss by spirting. The evaporation should not be commenced at the full heat of the water-bath. When dry, add a little hot water and more hydrobromic acid, and take down to dryness again. Without this second treatment some of the silver nitrate would escape conversion to bromide. Finally heat carefully over the bare flame till the substance begins to fuse.

Calculation and statement of results.—Calculate from the mean value obtained from Exercises III. and IV. what weight of silver combines with 1 of chlorine. From Exercise V. calculate what weight of bromine combines with this weight of silver.

For mode of statement of results and examples of results obtained, see end of Exercise IX.

Before proceeding to the analysis of potassium chloride (Exercise VIII.), two preliminary exercises must be performed.

The analysis of potassium chloride is effected by determining the chlorine in a given weight of the salt and subtracting the weight of the chlorine from that of the salt taken in order to find the weight of the potassium. The quantity of the chlorine needed is determined *from the weight of the silver needed to combine with it*, using the mean value obtained from Exercises III. and IV. The method we shall employ is a *volumetric* one, in which we estimate the chlorine by determining the volume of a silver-nitrate

D

solution of known strength which is just sufficient
to provide silver to combine with the chlorine contained in the weight of potassium chloride taken.
For carrying out this method we require to know
the percentage of silver contained in a silver nitrate
which forms the subject of the next exercise.

EXERCISE VI.—*Determination of the percentage of
silver in silver nitrate by the reduction of the
silver nitrate in hydrogen.*

The following are the *apparatus and substances
required:*—Crystals of *silver nitrate.* An *arrangement for heating the substance in a current of a gas,*
as described in Exercise III. A current of a reducing gas, either *coal-gas* or *hydrogen.* If coal-gas
be used it is well to let it bubble through a liquid,
in order that the rate of supply may be observed
and controlled. The materials for generating
hydrogen are zinc (the distilled zinc is the purest,
but it is expensive) and dilute pure sulphuric acid
(1 of strong acid to about 4 of water). The gas
may be purified by passing it through a solution
of potassium permanganate.

Method of conducting the experiment.—The reduction must not proceed too rapidly, as loss of
material might occur. The current of gas must
therefore be slow, and the reduction should be
carried out at as low a temperature as possible.
When no more brown fumes of oxides of nitrogen

arc formed the reduction is finished. The reaction is quickly effected.

Calculation and statement of results,—From the weight of silver nitrate taken and the weight of silver left, calculate how many parts of silver there are in 100 parts of silver nitrate. The following calculation will also be found useful. From Exercises III. and IV. we know the weight of silver which combines with weight 1 of chlorine ; calculate from Exercise VI. what weight of silver nitrate contains this weight of silver. The following is an example of results obtained. A pupil found that ·4765 silver nitrate gave ·3044 silver. The percentage of silver is therefore 63·88 per cent. found by the pupil, as against 63·50 per cent. given by standard determinations. Assuming the standard determination as absolutely correct, the experimental error of the pupil's determination is + ·38 per cent.

EXERCISE VII.—*Preparation of a solution of silver nitrate containing a known weight of the element silver per cb.c.*

The following are the *apparatus and substances required :* —Crystals of *silver nitrate, distilled water, solution of litmus, vaseline,* a 500-cb.c. *measuring-flask,* small *beaker,* small *funnel,* 50-cb.c. *pipette,* two *glass-stoppered bottles,* capable of holding at least 50 cb.c. each, *gummed labels.*

Method of preparing the solution.—Wash out the 500-cb.c. flask carefully with distilled water ; it need not be dried afterwards. Weigh out accurately about 8·5 grams of silver nitrate. The distilled water for dissolving the silver nitrate should be neutral, and must not contain more than a very small trace of chloride. To test the latter point add a drop of silver nitrate to some of the water. It should give at most a *very faint* turbidity. To test if neutral, pour one or two drops of blue (or, better, claret-coloured) solution of litmus into a porcelain dish and add about 25 cb.c. of the water. The colour should not change. The distilled water being such as is required, dissolve the silver nitrate in a small beaker, preferably one which has a lip for pouring. Place a small funnel in the neck of the 500-cb.c. measuring-flask and pour in the solution. A trace of vaseline put on the under side of the lip of the beaker will prevent the liquid from running down the outside. Rinse out the beaker several times into the flask. Make up the level in the flask so that the lower level of the meniscus touches the mark on the neck. Place the glass stopper, which must fit accurately, in the flask, and, holding this in its place, thoroughly mix the contents by inverting the flask once or twice. We have now a stock of silver nitrate of known strength, i.e. which contains a known weight of silver per cb.c. Transfer the solution to a clean and perfectly dry stoppered bottle. The drying must be effected by heating and blowing

air through, as described in Exercise II. Label
the bottle thus :

STANDARD SILVER NITRATE SOLUTION

1 cb.c. contains . . . silver = . . . chlorine
= . . . bromine
Name of pupil . . . . .
Date . . . . .

In the actual *titrations* a more dilute solution
made from this will be used.   It can be made up
from. time to time as required by withdrawing
50 cb.c. by means of a 50 cb.c. pipette, delivering
into a 500 cb.c. flask and making up to the mark on
the neck.   The pipette used must be clean *and dry*.
To dry it, blow air through from the bellows while
warming not too strongly in the luminous gas-
flame.   The pipette is held in a sloping position
with the delivery end downwards.   The delivery
end, in which the water collects, should not be
allowed to come near the flame.   In delivering the
solution from the pipette allow it to drain for about
a minute.   Then, when there is a length of liquid
in the narrow portion which does not form a drop
and fall, touch the side of the vessel.   About half
the liquid comes out ; the remainder of the liquid
should not be blown out, as the pipette is made to
deliver 50 cb.c., its contents being slightly greater.

Transfer the dilute silver-nitrate solution to a
clean, dry bottle, and label.

It generally happens that some pupils fall
behind others in a course of laboratory work.   If
it is desired that all should, as far as possible, work

the same experiments side by side, those who have fallen behind might omit the preparation of the standard silver nitrate solution, making up the dilute solution for themselves from some of the stock prepared by others.

EXERCISE VIII.—*Determination of the ratio (or proportion) in which potassium and chlorine are combined in potassium chloride.*

The following are the *apparatus and substances required*:—The *dilute silver nitrate solution* prepared in Exercise VII., *distilled water*, solution of *potassium chromate* free from chloride. To test if free from chloride acidify with pure dilute nitric acid and add a drop of silver nitrate solution. The liquid should remain clear. If a turbidity is produced, a purer specimen must be obtained, or the salt may be purified by re-crystallisation. To effect this, dissolve in the smallest quantity of water, with the aid of heat, in a small flask of about 150 cb.c. Cool under the tap, pour off the liquor from the deposited crystals, allow the crystals to drain thoroughly, dry them between folds of white blotting paper or filter paper, dissolve in water, and label ' Potassium Chromate—indicator.' We also require a 500 cb.c. *measuring flask*, small *funnel*, small *beaker*, *watch-glass*, a 25 cb.c. *pipette*, a *burette* and *burette-stand*, two or three *porcelain dishes* to hold 70 or 80 cb.c., one or two *glass stirring rods*, which are best made by sealing up the ends of glass tubing ; solid glass rods are apt to break the bottom of a beaker.

Method of conducting the experiment.—Weigh out accurately upon a watch-glass about ·35 gram of potassium chloride, and dissolve in 500 cb.c. of distilled water, which must be free from chloride and neutral to litmus. Wash out a burette with distilled water and allow it to drain. Then wash it out with from 1 cb.c. to 2 cb.c. of the dilute silver nitrate solution. Pour the dilute silver nitrate solution into the burette and see that it is filled to the end of the delivery jet. Take out 25 cb.c. of the potassium chloride solution with the 25 cb.c. pipette and deliver into a porcelain dish. If the distilled water was even slightly acid, add a few drops of pure carbonate of soda solution, free from chloride. Add a few drops of the potassium chromate solution. The same number of drops should be added in each titration. Read the lower level of the meniscus in the burette. If a float is used the reading of the burette is rendered much easier. Run the silver nitrate solution into the dish, keeping the liquid stirred by means of a glass rod. The silver solution will soon begin to produce a faint reddish tinge, due to the formation of silver chromate. This, however, is decomposed with the formation of silver chloride as long as any potassium chloride remains in the solution. When the amount of potassium chloride left is small the colour disappears slowly, an indication that the reaction is nearly completed. At length the addition of a drop of the silver solution produces a red spot, which on stirring diffuses a permanent faint, reddish tinge throughout the liquid. With the quantities prescribed this will be when *about* 25 cb.c. of the silver solution have been added. We have now added enough silver to combine with all the chlorine contained in the solution, and one drop over and above, which has formed the small quantity of red silver chromate which imparts a reddish tinge to the contents of the basin. The necessity for avoiding the presence of free acids throughout the operations arises from the fact that silver chromate is decomposed and dissolved by acids. Neglecting to take account of the drop in excess (equal to about ·05 cb.c.), we may say that the volume of silver nitrate solution added contains just so much silver as is required to combine with the chlorine contained in the 25 cb.c. of potassium chloride solution. Successive determinations are done with successive quantities of potassium chloride solution, until practice in hitting the exact point at which the reddish tinge appears enables the pupil to make successive determinations not differing by more than ·1 cb.c. The eye is assisted in noting the exact point of change by comparison with the contents of a dish containing potassium chromate solution and a soluble chloride to which some silver nitrate solution has been added, the soluble chloride being present in excess. A sheet of white paper placed beneath the dish is also of assistance. The colour change can be observed somewhat better by gaslight than by daylight.

Calculation and statement of results.—Multiply

the weight of chlorine corresponding to 1 cb.c. of the silver solution by the number of cb.c. used for 25 cb.c. of the potassium chloride solution. Multiply this number by 20 (i.e. 500 ÷ 25), and we obtain the weight of chlorine in the potassium chloride which was weighed out. Subtract this from the weight of the potassium chloride. The difference gives the weight of the potassium.

Calculate by simple proportion the weight of potassium which combines with weight 1 of chlorine.

EXERCISE IX.—*Determination of the ratio in which potassium and bromine are combined in potassium bromide.*

We require the same *apparatus and substances* as in the last exercises, except that potassium bromide is used instead of potassium chloride.

Method of conducting the experiment.—Weigh out accurately about ·5 gram of potassium bromide, and dissolve in 500 cb.c. of water. Proceed exactly as in the last exercise.

Calculation and statement of results.—The combining proportion of silver and bromine was determined in Exercise V. From the amount of silver used in the titration of the potassium bromide calculate the weight of bromine which it contains. The weight of potassium is obtained by subtracting the weight of bromine from that of the salt taken. Calculate the weight of bromine which is combined in potassium bromide with just so much potassium as has been found to unite with weight 1 of chlorine.

We have now obtained all the necessary data for a verification of the Law of Equivalent Proportions.

Example of results obtained in verification of the Law of Equivalent Proportions.—A pupil found that

1 part by weight of chlorine combines with
$\begin{cases} 3 \cdot 048 \text{ silver which combine with} \\ 2 \cdot 267 \text{ bromine} \\ 1 \cdot 1004 \text{ potassium which combine} \\ \quad \text{with } 2 \cdot 269 \text{ bromine} \end{cases}$

The bromine, instead of being exactly equal in each case, differs by 2 parts in 2,268 ; therefore the experimental error is ·07 per cent. The standard number for the percentage of silver in silver nitrate was employed by the pupil who obtained the above result.

The same results may also be calculated, using instead of weight 1 of chlorine, the weight of chlorine which standard determinations have shown to combine with weight 1 of hydrogen. The advantage of this method of calculation is that, by comparison with a table of *equivalent weights*, the degree of accuracy attained in each determination is readily calculated. The above example calculated on this basis comes out thus :—

35·37 chlorine combine with
$\begin{cases} 107 \cdot 80 \text{ silver (error } + \cdot 13 \text{ p. c.) which com-} \\ \quad \text{bine with } 80 \cdot 20 \text{ bromine (error } + 55 \text{ p. c.)} \\ 38 \cdot 92 \text{ potassium (error } - \cdot 28 \text{ p.c.) which com-} \\ \quad \text{bine with } 80 \cdot 26 \text{ bromine (} + \text{ error } 60 \text{ p. c.)} \end{cases}$

The *errors* recorded above represent the divergence
of the numbers from the standard determinations
of *equivalent weights*. It will generally be found
that the results obtained in this verification of the
Law of Equivalent Proportions has a smaller error
than some of the individual determinations.

*Note on early experiments relating to the Law of
Equivalent Proportions, and on Stas' determinations*
This law, which is sometimes called Richter's law,
was first enunciated as a relation between the
combining quantities of acids and bases. Richter's
results, published about 1793, were very rough.
Thus, his determinations of the ratio between
equivalent quantities of carbonic acid and lime
differs from the standard numbers as determined
by modern workers to the extent of 9 per cent.
He failed to convince the scientific world of his
day of the importance of his work. Cavendish
also published determinations of equivalent quantities of compounds. His numbers are much more
nearly accurate. When the law was applied to
the case of elements, the first results were again
very rough. Thus we find that Dalton ('New
System of Chemical Philosophy,' published 1808,
Part I., p. 219) gives numbers which show the ratio
of equivalent quantities of silver and sulphur to
be 7·69. The same numbers are given in Part II.
of Dalton's book published two years later. This
result is 14 per cent. greater than the ratio 6·75,
which is that of our present standard numbers.
The result calculated from the second experiment

cited in Chapter II. (in which silver was heated
with excess of sulphur) is 6·725, which differs from
the standard number by — ·4 per cent.   The
standard of accuracy for determination of equivalent quantities of the elements was greatly raised
by Berzelius.

As an example of modern work in which the
law of equivalent proportions is verified, we proceed
to quote numbers obtained by Stas.   They relate
to the same elements as those dealt with in the
exercises of the present chapter.   The weight of
silver is taken as unity instead of the weight of
chlorine, but the divergence from theoretical accuracy appears in the same way in both calculations,
viz. in the slightly different numbers for the two
determinations of bromine.

Stas found that :—

32,844·5 of chlorine combine with
{
100,000 of silver which combine with
74,080·5 of bromine
36,259·1 of potassium which combine
with 74,073·1 of bromine
}

The experimental error is here about ·01 per cent.,
or 1 part in 10,000.

For Stas' numbers quoted above, see ' Bull. de
l'Acad. Roy. Belg.,' 1860, No. 8, pp. 328 and 329, for
the weights of chlorine and of potassium equivalent
to 100,000 of silver ; and for the weight of bromine
equivalent to the above weights of silver and of
potassium, see ' Mém. de l'Acad. Roy. Belg.,' 1865,
vol. xxxv., pp. 171 and 172.

CHAPTER V

THE SECOND PART OF THE LAW OF DEFINITE AND OF CONSTANT PROPORTIONS

THE second part of this law states that the definite fixed proportion between the weights of two elements combining to form a particular substance is a *constant* proportion between the weights of those two elements present in chemical compounds containing also other elements.

For the verification of the first part of this law, see Chapter III.

EXERCISE X.—*Determination of the constancy of the ratio, or proportion, between the weights of potassium and chlorine in potassium chloride and in potassium chlorate.*

Having determined the proportion in which potassium and chlorine are combined in potassium chloride (Exercise VIII.), it may now be shown that these two elements are combined in the same proportion in potassium chlorate.

Apparatus and materials required.—A hard *glass tube* in which to heat the substance, closed at one end, and about 4 c.m. long, *retort stand* and *clamp, glass-wool, potassium chlorate*, the dilute *silver nitrate solution*, and other things required for titration, as in Exercise VIII.

Method of conducting the experiment.—Place about 1 gram finely powdered potassium chlorate, the weight of which need not be accurately known, in the dry glass tube. Plug the tube loosely about half-way up, or nearer to the open end, with glass-wool. Support the tube in the clamp of the retort stand near the open end of the tube, sloping it at an angle of about 15°. Heat the salt with the flame of a Bunsen burner, holding the burner in the hand and moving the flame to and fro at first so as not to crack the tube. If the salt after fusing shows a tendency to solidify, the heat must be increased ; the evolution of gas will then for a time become more rapid. When the oxygen has nearly ceased coming off, it may be advisable to turn the tube round in the clamp, so as to ensure that the flame shall get to all parts of the salt. When no more bubbles of gas come off, allow the tube to cool, and when quite cold weigh the tube with its contents. Place the tube in a basin of distilled water, which must be free from chloride, and extract the residue of salt in the tube by leaving in the hot water. Take out the tube, wash it inside and outside with a stream of water from the wash-bottle, allowing the washings to flow into the basin. Set the basin with its contents aside to cool. Wipe the glass tube with a cloth, and put it in a desiccator

to dry thoroughly. When dry weigh the empty
tube. The difference between this weighing and
that of the tube before extracting the salt gives
the weight of the salt extracted, in which we have
to determine the quantity of chlorine. Make the
volume of the solution up to 250 cb.c. and titrate
successive quantities of 25 cb.c. with the dilute
solution of silver nitrate used in Exercise VIII.
On comparing the results with those of Exercise
VIII., it will be found that for the same weight of
the salt in either case the same quantity of silver
nitrate is required ; hence each contains the same
quantity of chlorine.

For the purpose of this experiment it is not
necessary to have a standardised solution of silver
nitrate ; it will suffice to compare the quantity of
a solution of unknown strength required for a given
weight of a specimen of potassium chloride with
the amount required for the same weight of the
residue obtained after heating potassium chlorate.
This method of performing the experiment is
convenient if it is desired to verify the second part
of the Law of Definite and of Constant Proportions
before entering upon the Law of Equivalent Proportions. The following example shows the results
obtained by a pupil working in this way.

Statement of results.—A pupil found that a
certain weight of potassium chloride required
38·4 cb.c. of a solution of silver nitrate of unknown
strength. The same weight of residue left on
heating potassium chlorate required 38·2 cb.c.

As the amount of silver nitrate used is proportional to the amount of chlorine present, it follows that the weight of chlorine in equal weights of the two materials is the same to within 2 parts in 383, i.e. to ·5 per cent.

The verification of the law is, however, somewhat more rigorous if the strength of the silver nitrate, and the composition by weight of silver chloride are known, as will be the case if the pupil has conducted the preceding experiments. Assuming this knowledge, the proportions between potassium and chlorine in the two materials can be directly calculated. The above experimental result, if calculated out in this way, gives the proportion ·9 per cent. higher in the chlorate than in the chloride. The fact that the potassium is estimated *by difference* makes the experimental error come out higher than in the first calculation. The example of potassium chlorate and chloride serves to verify the law to the degree of accuracy required in a *first approximation*, but it appears that the chlorate loses a *small* quantity of chlorine on heating, and the method described above would not be suitable for determinations of the highest accuracy.

Note on Stas' determinations relating to the second part of the Law of Definite and of Constant Proportions.—It does not appear, as far as the present writer is aware, that the earlier chemists conducted experiments expressly designed to test whether the proportion between the weights of two elements which combine together to form a 'binary' compound is the same as the proportion between those two elements in a 'ternary' compound, i.e. in a compound in which the two elements are present along with a third. Ordinary analysis furnished some evidence in favour of this conclusion, but for the most part it seems rather to have been assumed than to have been proved. If the atomic theory be correct, it follows that the proportion by weight of two elements, A and B, present in a binary compound AB, will be the same as the proportion between the weights of those two elements in a ternary compound ABC ; or at any rate there will be a very simple relation between the quantities. Dalton's graphic symbols for ternary compounds, e.g. alcohol, show that he assumes this to be so. Stas, however, pointed out that it was important for the proper substantiation of the atomic theory that the point should be carefully tested. Having verified the fact that combining proportions are independent of conditions of pressure, temperature, &c. (by his analysis of ammonium chloride, *vide* Chapter III.), he proceeds to consider what further experiments are necessary to establish the invariability of the combining proportions of the elements. He says ('Mém. de l'Acad. Roy. Belg.,' vol. xxxv. 1865, p. 61): 'The constant composition of stable chemical compounds being admitted . . . it remains to be shown that in binary and ternary compounds, for example, having two elements in common, the elements

E

common to both are present in the same proportion by weight. Thus in two bodies, AB and ABC, the ratio of the weight of A to that of B should be exactly the same in AB as in ABC. The solution of the problem is independent of the ordinary operations of analysis ; it is sufficient to determine whether the ternary bodies can be reduced to binary without any fraction, however small, of one element common to both becoming free.'

To test this, Stas reduced by the action of sulphurous acid the chlorate, bromate, and iodate of silver to the condition of chloride, bromide, and iodide respectively. *No trace* of silver or of the other constituent was set free. To give an idea of the accuracy of this verification of the law of constant proportions, it should be mentioned that considerable quantities of the salt were used, e.g. in the case of the iodate about 70 grams. The accuracy is probably considerably greater than in the case of Stas' result relating to the law of equivalent proportions quoted in Chapter IV.

CHAPTER VI

THE LAW OF MULTIPLE PROPORTIONS

STATEMENT OF THE LAW.—*If there be more than one substance formed by the combination of two elements, then, taking the weight of one element as fixed in each substance, the weight of the other element in the second compound bears a simple proportion to the weight of that element in the first.*

TWO methods of verifying the law are given in this chapter. The first (Exercises XI., XII., and XIII.) is the more complete. In these exercises the proportions between chlorine and copper in each of the two chlorides of copper is determined. The second method (Exercise XIV.) is less complete, but may be adopted instead of the first method if time presses. The second method consists in determining the quantities of bromine in the two bromides of mercury. The quantity of mercury in each case is calculated by difference, and the proportions between mercury and bromine in the two salts are then worked out.

First method of verifying the law (Exercises XI., XII., and XIII.) *by the analysis of cuprous*

E 2

chloride ana cupric chloride.—We find how much silver chloride is formed in each case and how much subsulphide of copper. If the cupric chloride gives twice as much silver chloride as does the cuprous chloride for the same weight of subsulphide of copper, then it is evident that for the same quantity of copper the amount of chlorine in the cupric salt is *twice* the quantity in the cuprous salt. It is not *necessary* to know the composition by weight of either the subsulphide of copper or of the chloride of silver for the purpose of verifying the law.　　　.

The following are the *apparatus and materials required* for this method of verifying the law :— *Cuprous chloride, cupric oxide* ; strong, pure *hydrochloric* and *nitric acids,* large *beaker, syphon-tube,* solution of *silver nitrate,* flowers of *sulphur, vaseline,* supply of *hydrogen* or *coal gas,* and a double set of the following apparatus, if, as is recommended, the two determinations be carried on side by side, viz. a 250 cb.c. *measuring flask,* 50 cb.c. *pipette, Rose crucible* (if this be not available a clay tobacco pipe fitted into an ordinary porcelain crucible may, according to Fresenius, be substituted), a *porcelain crucible* and *lid,* an accurately cut *funnel* with angle of 60°, *watch-glass, beaker, iron tripod, pipe-clay triangle, sand tray, evaporating basin,* holding more than 250 cb.c., *water-bath, desiccator* containing lime or caustic potash, hollow *glass-rod,* porous or *biscuit porcelain,* packet of *Swedish filter papers,* stoutest *platinum wire, camel-hair brush,* or, better, a *stiff black feather, glazed paper, filter stand, drying oven* with *thermometer, wash bottle, mortar.*

EXERCISE XI.—*Preparation of the two chlorides of copper in a state of purity.*

Cuprous chloride.—(The pupil should be acquainted with the mode of formation of this salt.) Take the salt as obtained from the dealers (which is *not* pure), dissolve it in a little strong, pure hydrochloric acid, and pour the clear solution into a large beaker of distilled water. The cuprous chloride is thrown down as a white precipitate, which quickly settles to the bottom of the beaker. Syphon off the bluish solution as quickly as possible ; the loss of some of the precipitate does not matter. Fill the beaker again with distilled water, stir up, allow to settle, and syphon off again. Repeat the operation a third time, to ensure that the washings are free from cupric salt. The pure white cuprous chloride which remains at the bottom of the beaker would readily oxidise if dried. No attempt, therefore, is made to dry and weigh the substance, it being sufficient for our purpose to determine the ratio of copper to chlorine in an unknown weight of the salt. Immediately after the last washing add strong pure nitric acid, free from hydrochloric acid, little by little until the salt has dissolved. Make up the volume of the liquid to 250 cb.c., and label the flask in which the solution is contained. The next step is to deter-

mine the quantity of cuprous sulphide and of silver
chloride respectively which can be obtained from two
equal measures (of 50 cb.c.) drawn from the flask
with a pipette.   Should anything occur to prevent
the successful completion of either determination,
another portion of 50 cb.c. could be withdrawn from
the flask and the determination repeated.   Cupric
chloride is analysed in the same manner ; the salt
is dissolved in water, without being dried or
weighed, the solution made up to 250 cb.c. ; two
measures of 50 cb.c. each are taken, in one of which
the copper is determined as subsulphide, and in the
other the chlorine is determined as silver chloride.
Economy of time is effected by conducting side by
side the two chlorine determinations, and afterwards
the two copper determinations.

Cupric chloride.—To prepare pure cupric chloride, powder finely some pure black oxide of
copper, and roast it in a porcelain dish over a Rose
burner.   Treat an excess of the oxide with strong,
pure hydrochloric acid with the aid of heat ; pour
off from the excess of oxide, and cool the hot
solution rapidly, placing the boiling tube or
narrow beaker in which it is contained under water
running from a tap, and keeping the contents of
the vessel rapidly stirred with a glass rod, when
the cupric chloride crystallises out.   Pour off the
liquid as completely as possible, and allow the
crystals to drain.   Spread the crystals upon a
piece of porous porcelain, and allow them to dry
in a desiccator over quicklime or pieces of caustic

potash. Dissolve the substance in water. If any matter of a bluish-white colour remain undissolved, filter through Swedish filter paper. Make up the liquid to 250 cb.c. Separate portions of 50 cb.c., each can then be used for analysis.

EXERCISES XII. AND XIII.— *The analysis of the two chlorides of copper.*

For the *copper determination* (Exercise XII.) remove 50 cb.c. of the solution by means of a pipette and deliver it into a small evaporating basin. Concentrate the solution on a water-bath to small bulk. Weigh a Rose crucible with lid, finish the evaporation to dryness in the Rose crucible, which if too small for the rings of the water-bath may be conveniently supported by a pipe-clay triangle. When quite dry add flowers of sulphur, place the perforated lid upon the crucible, connect up with a slow stream of hydrogen, and after having displaced the air from the crucible by means of the hydrogen, heat the crucible gently at first and afterwards more strongly with the flame of the Bunsen burner, and finally with the foot-blowpipe, using not quite the full blast, and discontinuing the heating five minutes after the flame of sulphurous acid or the fumes of sulphur have ceased to be visible round the lid of the crucible. Allow to cool in the current of hydrogen. Examine the contents of the crucible, which should contain a dark-coloured, shining, crystalline mass of subsulphide of copper with no visible sulphur and no red colour, which would indicate reduced copper. Weigh the crucible and lid with the contents.

For the *chlorine determination* (Exercise XIII.) take up 50 cb.c. of the solution of copper salt in a pipette and deliver into a small beaker. In the case of the cupric chloride add a few drops of nitric acid. Place the beaker on the sand-tray, cover the mouth of the beaker with a watch-glass, and heat to boiling point. Add solution of silver nitrate. Boil for a few minutes, till the precipitate collects in flocks. Allow the precipitate to sub-side. Add a drop more solution of silver nitrate. If no fresh precipitate is produced, all the chlorine has been thrown down as silver chloride.

It is important not to add a great excess of a reagent for the precipitation of an insoluble com-pound ; on the other hand, if too little of the re-agent be added at first, time is lost by having to add a second dose. Whenever the pupil can form an estimate of the weight of a reagent required for precipitation, he should use such a volume of the solution of the reagent as contains a weight *slightly* greater than is absolutely necessary. The reagents in a laboratory are made up roughly to a certain strength (so many grams to the Winchester-bottle of water), and the pupil should be acquainted with the strength of the solutions and be able to calculate approximately what weight of each re-agent is contained in 1 cb.c. of the solution. It is a good plan to write on the label of each reagent bottle the number of grams per Winchester-bottle used in making up the solution. Dilute acids should be marked 1 : 3, 1 : 5, &c., according to the proportion of acid to water which has been used.

Having made sure that all the chlorine is pre-cipitated as silver chloride, fit carefully a small Swedish filter paper to a well-shaped glass funnel. In order to hasten the filtering it is well to attach to the funnel a glass tube provided with one bend in order to create a suction. Before commencing the filtration it is well to ascertain, by pouring water on the filter paper, that the paper is strong enough to stand the suction: The use of filtering pumps in quantitative experiments should not be attempted until the pupil has had experience in the management of filtrations. Very slightly grease with vaseline a small portion of the underside of the rim of the beaker, and pour the liquid down a glass rod (made from hollow tubing) into the funnel to about two-thirds of the height of the filter paper. The liquid should be poured so as to fall on the side of the funnel, not into the bottom of the cone. Have a perfectly clean beaker to catch the filtrate, since, although it is intended to throw away the filtrate, yet it is important that if through mishap some of the precipitate should come through the filter paper, this should be recoverable by filtering again. When the first charge of liquid has run through, fill the funnel two-thirds full again, and repeat the operation till the liquid above the pre-cipitate in the beaker has all been poured off.

Then pour hot distilled water on the precipitate in the beaker and boil up for one or two minutes. Allow the precipitate to subside and filter again. After sufficiently washing the precipitate in this way (by 'decantation'), bring the precipitate carefully on the filter paper with the aid of the glass rod. The last portions must be removed by the aid of a jet of water from the nozzle of a wash-bottle. Having brought all the precipitate on to the filter paper, wash further with a stream of hot water from the wash-bottle until the washings appear to be pure water, giving no reaction for copper, for an acid, or for silver. Having washed the precipitate, wet a piece of filter paper, and place it over the mouth of the funnel which contains the precipitate. Press the edges of the filter paper to the sides of the funnel and tear so as to leave a circular portion of the filter paper tightly stretched over the mouth of the funnel. Place the funnel in a drying-oven, or 'air-bath,' heated to a temperature not much exceeding 100° C., and leave till dry. The temperature of the air-bath should be ascertained from time to time by reading the thermometer, which passes through a cork fitted in the opening at the top. While the precipitate is drying, weigh a porcelain crucible and lid. *To determine the weight of the ash of the filter paper*, take four to six filter papers of the same packet, wrap them tightly in a small bundle and hold by two turns of *stout* platinum wire. Set fire to the bundle and burn so that the ash will fall into the crucible. In order to ensure this see that the burning paper is not exposed to draughts. When the bundle of paper has ceased to glow tap the wire, if necessary, so as to make the bundle fall into the crucible. Strongly heat the crucible, placing the lid loosely on, till there is no black carbonaceous material left with the ash. Place the crucible in a desiccator, allow to cool, and weigh. Divide the total weight of the ash of four, or six, filter papers, by the number of papers used. The quotient will be the weight of ash of one filter paper of the packet. This weight should be written on the band which holds the papers, and should also be entered in the pupil's note-book. It is not advisable to rely upon the weight of ash printed on the wrapper. Having determined its weight, the ash may be thrown out of the crucible, and after testing again the weight of the crucible and lid, the crucible may be used in the weighing of the precipitate.

Take the funnel from the oven. If no steam is escaping, if the paper does not stick to the glass, and if the precipitate is loose and crumbling, the substance is dry, and may safely be removed from the filter paper. The *removal of a dried precipitate* from the filter paper to the crucible is an important piece of manipulation in quantitative analysis. A beginner should *rehearse* the manipulation before conducting the quantitative analysis. In the previous experiments in which silver chloride has been dealt with the whole operation was ꞏ ʰout transference, in one vessel. It may probably be assumed that transference is always accompanied by loss. With the small quantities used in ordinary chemical analyses it is, however, possible by proper methods of manipulation to lose so little that the loss is inappreciable with the balances employed. If the loss is too small to affect the balance it is the same, so far as numerical results are concerned, as if there were no loss at all. The removal of a precipitate from filter paper to crucible is a perfectly legitimate operation provided we are dealing with small quantities, and can only use a balance of ordinary sensitiveness. In researches such as those of Stas, in which large quantities and very delicate balances are used, the processes are arranged so as to avoid transference of material.

The crucible must be placed upon a piece of dark-coloured glazed paper or upon a large clock-glass, so that if any substance be spilt it may be recovered. Most of the precipitate on the filter paper can be made to fall into the crucible by scraping *gently* with a loop of stout platinum wire. The remainder must be gently brushed off with a stiff black feather, taking care not to brush off any of the surface of the filter paper. The paper, having been freed as much as possible from the precipitate, is rolled up into a small tight scroll and burnt over the crucible, being held by one turn of stoutest platinum wire. It is convenient to use a piece of platinum wire sufficiently long to fix one end on the stem of an inverted glass funnel, which

serves for a holder, while the paper is burning.
Care must be taken to burn the carbonaceous
matter as completely as possible, but the residue
should not be very strongly heated on the wire, as
the reduced silver would be apt to adhere to the
platinum instead of remaining in grains and falling
into the crucible with the little roll of ash. We
have now the silver chloride in the crucible. The
silver there is just so much of the silver in the
silver nitrate solution as was needed to seize upon
and throw down in the form of silver chloride all
the chlorine which was originally combined with
the copper in the 50 cb.c. of solution of the copper
salt. The rest of the silver remained in solution,
and was washed away in the filtering process. *All*
the chlorine in the 50 cb.c. of the solution was
thrown down as silver chloride ; but in burning the
filter paper the organic matter will have reduced
to metallic silver so much of the chloride as adhered
to the paper. It follows that the material in the
crucible weighs less than the silver chloride
precipitated, owing to the fact that a small quantity
of silver has been deprived of its chlorine. We
have verified by previous experiments the fact that
the proportion by weight in which silver and
chlorine combine does not depend upon the nature
of the reaction by which the combination is brought
about. Therefore by causing the small quantity of
reduced silver to combine again with chlorine we
shall obtain the true weight of the silver chloride
which was precipitated. To effect this, moisten

the contents of the crucible with one drop of pure
strong hydrochloric acid, heat very gently at first
and afterwards more strongly, but use a small
flame of the Bunsen burner, and move the flame
about slowly so that no part becomes very highly
heated. The lid is not placed upon the crucible
during this operation, which can be watched until
the substance begins to melt round the edges of
the crucible, when the lamp should be withdrawn.
Place the crucible in a desiccator, replace the lid,
allow to cool, and weigh. From this weight subtract the weight of crucible and lid *plus* the weight
of the ash of one filter paper. The difference is
the weight of the silver chloride formed from the
chlorine contained in 50 cb.c. of the solution of
copper salt.

*Example of the calculation and mode of statement
of results.*—The manner of stating the verification
of the Law of Multiple Proportions may be shown
by the following example from determinations
conducted when rehearsing the experiment to
ascertain if it were suitable to be tried for teaching
purposes. Results obtained by pupils are given
further on.

Equal portions of cupric chloride solution gave ·3139 grams copper
   subsulphide and 1·1141 silver chloride
Equal portions of cuprous chloride solution gave ·1872 grams
   copper subsulphide and ·3351 silver chloride
Therefore, so much cupric chloride as would give ·1872 grams
   copper subsulphide would give ·6644 silver chloride

Now $\dfrac{6644}{3351} = 1\cdot983$ (found) as against 2·000 calculated on the assumption of the Law of Multiple Proportions. The experimental error therefore is 17 parts in 2000, = − ·85 per cent. The error appears with the minus sign. If we had taken the weight of chloride as constant, and had calculated the amounts of sulphide, the error would have been the same in amount but opposite in sign. It is a matter of convention therefore, but one or other system of calculation must be adhered to, and the sign (+ tive or − tive) stated, if the results of a number of determinations are to be compared.

For teaching purposes it may be thought convenient to have some check upon the experiments during the course of the work. For this purpose standard numbers may be taken for the percentage composition of the subsulphide of copper, and the chloride of silver, and of the two chlorides of copper, and the results obtained by the pupil may be compared with these standard numbers as the exercises proceed. The two following sets of examples of results obtained by two pupils show this mode of statement.

The *first pupil* obtained the following results :—

CUPRIC CHLORIDE

	Found	Calculated from standard numbers	Experimental error calculated for chlorine
Copper .	48·00 p. c.	47·18 p. c.	+ ·82 p. c. of the
Chlorine .	52·00 ,,	52·82 ,,	total quantity of the salt used

CUPROUS CHLORIDE

	Found	Calculated from standard numbers	Experimental error calculated for chlorine
Copper .	64·68 p. c.	64·11 p. c.	+ ·57 p. c. of the
Chlorine .	35·32 ,,	35·89 ,,	total quantity of the salt used

Therefore, by simple proportion, in cuprous chloride, the ratio of copper to chlorine was found to be as 48 : 26·2, against 48 : 52 in cupric chloride.

Now $\dfrac{52}{26\cdot2} = 1\cdot985$ (found) as against 2·000 required by the Law of Multiple Proportion, experimental error − ·75 per cent.

The *second pupil* obtained the following results :—

CUPRIC CHLORIDE

	Found	Calculated from standard numbers	Experimental error calculated for chlorine
Copper .	48·04 p. c.	47·18 p. c.	+ ·86 p. c. of the
Chlorine .	51·96 ,,	52·82 ,,	total quantity of the salt used

CUPROUS CHLORIDE

	Found	Calculated from standard numbers	Experimental error calculated for chlorine
Copper .	64·46 p. c.	64·11 p. c.	+ ·35 p. c. of the
Chlorine .	35·54 ,,	35·89 ,,	total quantity of the salt used

Therefore, by simple proportion, in cuprous chloride the ratio of copper to chlorine was found to be 48·04 : 26·48, against 48·04 : 51·96 in cupric chloride.

Now $\dfrac{51·96}{26·48} = 1·962$ (found) as against 2·000 required by the Law of Multiple Proportions, experimental error −1·9 per cent.

Second method of verifying the Law of Multiple Proportions (alternative to that of analysis of chlorides of copper).

Exercise XIV.—(*Alternative to Exercises XI.,
XII., and XIII.*), *the determination of the percentage of bromine in the two bromides of mercury.*

By this exercise the Law of Multiple Proportions can be verified on the assumption that the
salts known as mercurous bromide and mercuric
bromide are composed only of the two elements
mercury and bromine. The pupil should be acquainted with the methods of preparation of the
salts. The exercise is intended to furnish a shorter
method of verifying the law; the salts are not
prepared by the pupil but are obtained direct from
the dealers, by whom they are supplied in a fairly
high state of purity. No attempt is made to
determine directly the quantity of mercury; it is
arrived at by subtracting the weight of the bromine
from the weight of the salt. The method of
determining the bromine volumetrically has been
employed in Exercise IX., and the composition by
weight of silver bromide was determined in Exercise V. Strictly speaking, the pupil should make
use of the numbers obtained by himself in Exercise V. when calculating the results of the present
exercise. In the following account of an experiment actually performed, the determinations were
carried out by a practised chemist, not by a pupil.
Standard numbers were assumed for the composition of silver bromide.

Account of an actual experiment, as conducted.—
A strong solution of caustic potash containing

F

about 20 per cent. was used for the decomposition
of the salts.   The sample was free from admixture
of chloride. *Mercurous bromide*, 1 gram, was weighed
out and covered with 20 cb.c. of the caustic potash
solution in a small porcelain dish, and warmed on
the water-bath for half an hour.   The decomposition, however, was almost immediate.   Water
was added to dilute further the caustic alkali, the
liquid was filtered, and the filtrate made up to a
volume of 250 cb.c.   The mercury is all left on the
filter paper as oxide, &c., and all the bromine is in
the filtrate as potassium bromide.   In the filtrate
the bromine was determined volumetrically by
exactly neutralising 20 cb.c. with pure nitric acid,
free from chloride, and titrating with a solution of
silver nitrate of known strength, using potassium
chromate as an indicator (see Exercises VIII. and
IX.).   Two determinations gave the weight of
bromine, calculated for the whole 250 cb.c. as

1st determination	.	.	.	·283 grams bromine
2nd ,,		.	.	·284 ,,
Mean	.	.	.	.	·2835 ,,

The weight of mercurous bromide was exactly
1 gram, therefore the composition of the salt is :—

	Found	Calculated from standard numbers
Bromine .	. 28·35 p. c.	28·57 p. c.
Mercury .	. 71·65 ,,	exp. error − ·22 p. c.

In the analysis of the *mercuric bromide*, 1 gram of
the salt was weighed out and dissolved in about
100 cb.c. of boiling water.   Twenty cb.c. of the

solution of caustic potash was added, the mercury being thrown down from solution in the form of the red oxide. The liquid was then filtered from the precipitated oxide, Swedish filter paper being used. If a small quantity of the precipitate should have passed through the filter paper the titration could still have been proceeded with, as the precipitate quickly settles to the bottom of the measuring flask. The filtrate was neutralised as in the last experiment, the volume made up to 250 cb.c., and 20 cb.c. were titrated with the same solution of silver nitrate as was used for the mercurous bromide.

The weight of bromine found in three determinations was :—

1st determination	.	.	.	·446 grams bromine
2nd ,,		.	.	·443 ,,
3rd ,,		.	.	·437 ,,
Mean .	.	.	.	·442 ,,

The weight of the mercuric bromide was exactly 1 gram, therefore the composition of the salt is :—

	Found		Calculated from standard numbers
Bromine .	.	. 44·2 p. c.	44·4 p. c.
Mercury .	.	. 55·8 p. c.	exp. error − ·20 p. c.

Mode of stating the above results as a verification of the Law of Multiple Proportions.

In mercuric bromide 55·8 mercury are combined with 44·2 bromine
In mercurous ,, 71·65 ,, ,, 28·35 ,,
Therefore, in mercuric bromine 71·65 mercury are combined with 56·75 bromine

Now $\dfrac{5675}{2835} = 2\cdot002$ (found) as against $2\cdot000$ calculated on the assumption of the Law of Multiple Proportions. The experimental error is therefore $+ \cdot1$ per cent.

Note upon early experiments relating to the Law of Multiple Proportions and upon Stas' criticisms of these experiments.

The results obtained by chemists in the earlier part of the century were for the most part less nearly exact than those given above.

Dalton, ' New System of Chemical Philosophy,' p. 318, quotes in support of the law the following values obtained by Sir Humphry Davy for the composition by weight of two oxides of nitrogen :—

	Nitrogen	Oxygen
Nitric oxide . . .	5·3	7
Nitrous oxide . . .	11·46	7

Now $\dfrac{11\cdot46}{5\cdot3} = 2\cdot16$ (found) as against $2\cdot00$ required by the Law of Multiple Proportions, the experimental error is therefore $+ 8$ per cent. Dalton's own results for the hydrocarbons and for the oxides of carbon are even further from the numbers required by the law.

The results obtained by Berzelius are more nearly exact. Thus, e.g., he found the ratio between the quantities of the base in the sub-arsenite and arsenite of lead to be as

$$1 : 1\cdot974 \qquad \text{exp. error} \sim 1\cdot3 \text{ p. c.}$$

and he found the oxygen in chromic oxide and chromic acid to be as

$$1 : 2·062 \qquad \text{exp. error} + 3·1 \text{ p. c.}$$

Probably chemists were influenced in their acceptance of the Law of Multiple Proportions upon results so rough by the circumstance that it suggested a ready explanation (that of chemical atoms) of the better established law of definite, fixed proportions. Stas remarks in the introduction to his ' Nouvelles Recherches sur les Lois des Proportions Chimiques,' that chemists and physicists have long been in the habit of crediting the existence of a simple mathematical relation whenever phenomena present an appearance of regularity. This prejudice, he says, leads them to ascribe observed deviations *wholly* to experimental error, and he cites the case of the hypothesis, called ' Prout's,' that the weights of all chemical atoms are exact multiples of the weight of the hydrogen atom. Stas verified the Laws of Definite, of Constant, and of Equivalent Proportions, as well as the Law of Conservation of Mass, but did not examine the Law of Multiple Proportions considered as *Loi mathématique* and not merely as *Loi limitée.* Although the discovery of the Law of Multiple Proportions may have suggested Dalton's atomic theory, Stas considers (*loc. cit.*, p. 60 of the ' Mém. de l'Acad. Roy. Belg., vol. xxxv., 1865) that chemists have relied more upon the definite and constant proportions of chemical combination as evidence of the truth of Dalton's theory.

CHAPTER VII

THE LAW OF SIMPLE VOLUMETRIC PROPORTIONS IN THE CHEMICAL REACTIONS OF GASES

UPON the molecules of gases.—The development of the modern theory of chemical atoms and chemical molecules from Dalton's atomic theory commenced with the discovery of a relation between Dalton's 'chemical atoms' and the 'molecules' of gases. The molecule of a gas is defined by physicists as 'a small mass of matter, the parts of which do not part company during the excursions which the molecule makes when the body to which it belongs becomes hot' (Clerk Maxwell, 'Theory of Heat,' 6th edition, p. 305). This physical definition is independent of chemical considerations. It appears from purely physical facts that the number of molecules in a given volume is the same for all gases at the same temperature and pressure (*loc. cit.*, pp. 301–317).

The researches of Gay-Lussac ('Mém. Soc. d'Arcueil,' vol. ii., 1809, reprinted by the Alembic Club, 'Reprints,' No. 4) showed, within fairly wide limits of experimental error, that the volumes of gases which react chemically together bear a simple ratio, or proportion, to one another. Chemists were at first inclined to infer from this that the chemical 'ultimate particles' or 'atoms' of gases are identical with the physical 'molecule.' Further examination of the matter showed that in the case of most chemical elements in the gaseous state, this could not be so. We shall return to this point after formally stating Gay-Lussac's 'Laws of Volume' in a condensed form ; merely stating here that the conclusion arrived at has been that the molecule of a compound gas is identical with Dalton's ultimate particle or reacting unit, but that in the case of elementary gases the molecule may contain one *or more* chemical atoms. Most of the common elementary gases (e.g. hydrogen, oxygen, nitrogen, and chlorine) have two atoms in the molecule. The following is a condensed statement of the Laws of Simple Volumetric Proportion : *The volume of an element in the gaseous state bears a simple proportion to the volume of the compound gas of which it is a constituent.* From this statement it would follow as a corollary :—

Corollary I.—The volumes of the gaseous constituents of a compound gas bear a simple proportion to one another.

It may be further deduced (assuming the laws of definite and of constant proportions) that the volumes of the combining or equivalent weights of gaseous substances bear a simple proportion to one another, or, as it may be stated :—

Corollary II. — The relative densities of gases stand in a simple ratio to their chemical equivalents.

This relation was experimentally observed about the same time that the laws of volumetric proportions were discovered. In the case of the elementary gases the corollary may be expressed by saying that the relative densities stand in a simple ratio to the relative weights of the atoms.

Having now explained the connection between Gay-Lussac's volumetric laws and the gravimetric laws which formed the foundation of Dalton's theory, we pass on to describe an experimental verification of the law of simple volumetric proportion.

EXERCISE XV. — *The determination of the volume of nitrogen obtained by the decomposition of a known volume of ammonia.*[1]

The reaction employed is the decomposition of ammonia gas, contained in a graduated tube, by a solution of sodium hypobromite, the hydrogen of the ammonia being oxidised and the nitrogen set free. The pupil should acquaint himself with the preparation and properties of the hypobromites (and hypochlorites), and with the evidence showing

[1] This experiment is described by Professor Ramsay in his *Experimental Proofs of Chemical Theory*, from which it is taken with his permission. The directions for performing the experiment have been to some extent modified in accordance with the present writer's experience.

that ammonia consists wholly of nitrogen and hydrogen.

Apparatus and materials required. — Strong solution of *ammonia* (' liq. amm. fort.'), *bromine* (not bromine-water), solid *caustic soda*, dilute *acetic acid, retort stand* with rings and clamp, *sand tray*, two *flasks, wide-mouthed bottle*, or Woulff's bottle, three *corks*, or rubber stoppers, *glass tubing* and connecting *rubber tubing*, graduated *gas-measuring tube* of 50 to 100 cb.c. capacity, not too wide to close with the thumb, and the narrower the better a *basin* in which to invert the tube (a small pudding-basin or a clean glass mortar is more convenient than the ordinary shallow evaporating basin), a *thermometer*, a large and deep vessel, or a *pneumatic trough* deep enough to immerse the graduated tube.

Before proceeding to the actual experiment, it is well that the pupil should rehearse the operations of closing the tube, placing under water, shaking the tube, &c., in order to make sure that he is master of the manipulation. A dish containing water to which a little acetic acid has been added should be at hand when the experiment is being conducted. After the hand has been immersed in the caustic soda solution it should be immediately and thoroughly washed in the basin.

Preparation of sodium hypobromide.—Make up a strong solution of caustic soda in a flask. In order to judge of the volume which will be required, pour into the pudding-basin (or glass mortar) as much water as will cover the thumb to the depth of an inch when the hand is put into the water in the position required for inverting the tube in the experiment. The volume of the caustic soda solution required will be from 50 to 100 cb.c. more than this. The flask should be corked while the solution is proceeding, and the flask should be cooled under the tap before adding the bromine. Add the bromine gradually to the cooled solution in the draught chamber, agitating the flask till the heavy red liquid has dissolved in the soda solution, and then adding more bromine. The temperature may be kept low, if necessary, by putting the flask under the tap. The formation of the hypobromite is accompanied by change to a greenish-yellow colour. Ten cb.c. of bromine may be used to 300 cb.c. of caustic soda solution. If 50 cb.c. of the solution enter the tube during the reaction, the amount of bromine (in the form of hypobromite) entering the tube would be largely in excess of the amount used up in the decomposition of 100 cb.c. of ammonia gas. A large excess of hypobromite greatly hastens the reaction, whereas, if the solution of hypobromite be a weak one, the last part of the reaction goes very slowly and the experiment cannot be completed within a reasonable time.

To fill the tube with ammonia gas, warm gently some strong ammonia solution in a small flask fitted with a one-way cork. The gas should be passed through a dry flask or a Woulff's bottle, in

which may be put some cotton wool to retain drops
of moisture, or lumps of quick-lime may be employed for the same purpose. Collect the gas by
upward displacement in the graduated tube. The
delivery tube should reach nearly to the top of the
graduated tube, which must be dry. After use in
the experiment the tube should be dried, before
using again, by warming and blowing air through
from the bellows. It is convenient, however, to
have more than one tube and to start two or three
experiments without waiting for the completion of
the first. The precautions taken as to dryness are
designed to exclude drops of moisture, which would
hold in solution relatively· large quantities of
ammonia gas and would vitiate the results. As
the tube is open to the air the gas will not, pre-
sumably, be *dry* in the more precise sense of the
term, but will contain vapour of water. The stopper
of a bottle of hydrochloric acid brought near to,
and a little above, the mouth of the graduated tube
will show by forming white fumes when the tube
is practically filled with ammonia gas. The
evolution of gas should be allowed to continue for
some time afterwards. Observe the temperature
of a thermometer suspended near the tube. Finally
raise the graduated tube very slowly, so that the
entering gas may fill the space previously occupied
by the narrow delivery tube. The tube should be
held in a twist of paper, or a glove should be worn,
lest the heat of the hand should expand the gas.
Close the mouth of the tube firmly with the thumb,

and plunge the tube mouth downwards to the
bottom of the basin which has been filled to a
sufficient depth with the solution of caustic soda
and hypobromite. Remove the thumb, keeping
the tube upright. The liquid rushes into the tube
and effervescence takes place. Clamp the tube in
the upright position ; watch and make notes of the
progress of the reaction. The large excess of
caustic soda retards the absorption of ammonia by
the liquid. After the first violence of the reaction
is over, most of the chemical change appears to
take place near the surface of separation between
the liquid and the gas. If the liquid be de-
colorised to a small depth, this indicates that the
hypobromite has been deoxidised in this part of
the tube. In this case the action mainly takes
place at the upper surface of the coloured liquid,
as may be seen by noticing the level from which
the bubbles of (nitrogen) gas rise. The rise or fall
of the level of the liquid at any particular time
depends upon whether the absorption of ammonia
or the evolution of nitrogen is proceeding the more
rapidly. The liquid may, for instance, rise two-
thirds or more up the tube and finally sink to one-
half way up. When the action becomes slow the
tube may be shaken from side to side, taking care
to keep the bottom of the tube safely under the
liquid in the basin. If time presses, the tube may
even be closed with the thumb and the contained
liquid and gas be agitated together. If this be
done the liquid should be agitated in such a way

that the thumb remains covered by it. When no more bubbles of gas come off, close the tube with the thumb under the surface of the liquid in the basin and transfer the tube to the large and deep vessel of water, the temperature of which should be brought to an equality with the temperature registered by the thermometer which was suspended near the graduated tube when the ammonia was being collected. After the measuring tube has remained in the water for five or ten minutes, raise it until the level of the water is the same inside the tube and outside, and read off the volume of the gas to the bottom of the meniscus. The tube should be held by a clip or ·by a twist of paper, not by the unprotected hand, lest the heat of the hand should cause the gas to expand.

Example of the calculation and mode of statement of results.—A senior student obtained the following results from two experiments which were completed in 2½ hours, using two tubes of different sizes. The first tube had been dried and was being used again for a third experiment, which was not quite completed at the end of the 2½ hours. The apparatus had been got together and fitted up on the previous day. *First experiment*, done with a rather wide measuring-tube graduated in ½ cubic centimetres. The divisions not being continued to the open end of the tube, the distance from the last division to the end of the tube was measured, and found to be equal in length to 15 of the cubic centimetre divisions. The volume of the tube was therefore

taken to be 115 cb.c.   The results were as follows :—

Volume of ammonia. . . .	115 cb.c.
,, nitrogen	58 ,,

Ratio 58 : 115 = 1 : 1·983 (found) as against
2·000 which is the nearest simple ratio.   The
experimental error is therefore −·85 per cent.
The minus sign indicates that the volume of
nitrogen was found to be *more* than one half the
volume of the ammonia.

Second experiment, done with a narrow eudio-
meter tube graduated in millimetres.   The volume
of the ungraduated part at the upper end of the
tube was known to be equal to 3 of the m.m.
divisions.   The number of m.m. divisions was 250,
and the length from the last m.m. division-mark to
the open end of the tube was found to be 57·5 m.m.
The volume of the tube was therefore reckoned
to be 3 + 250 + 57·5 = 310·5.   For the purpose of
the experiment it is not necessary to know the
actual volume of one division.   The results were as
follows :—

Volume of ammonia	310·5
,, nitrogen	154

Ratio 154 : 310·5 = 1 : 2·016.

The experimental error is therefore + ·8 per
cent., the volume of nitrogen being found to be
slightly less than one half the volume of ammonia.

*Note upon early experiments relating to the law
of simple volumetric ratios, or proportions.*—Gay-
Lussac, calculating from H. Davy's analyses, finds

the volumetric proportions in the oxides of nitrogen to be as follows. The numbers given in the last column show how fa rthese results depart from the numbers required to give the nearest simple ratios.

	Nitrogen	Oxygen	Error
Nitrous oxide . .	100	49·5	1 p. c.
Nitric oxide . . .	100	108·5	8·5 ,,
Nitrogen peroxide . .	100	204·7	2·35 ,,

Dalton's nearest result for the volumetric composition of water vapour was :—

	Oxygen	Hydrogen	Error
Water (vapour) . .	100	197	1·5 p. c.

These are examples of the results upon which the law was accepted by chemists.

Of recent years the volumetric composition of water has been carefully determined by Scott, who finds the ratio to be 1·9965, which confirms the law to within ·18 per cent.

Upon the relation of the (physical) molecule of nitrogen gas to the (chemical) atom of the element nitrogen.—Exercise XV. not only confirms the law of simple volumetric proportions but affords evidence as to the relation between the molecule of nitrogen gas and the chemical atom of the element. Equal volumes of nitrogen and ammonia contain,

according to the physical theory, an equal number
of molecules.   It was found in Exercise XV. that
ammonia when decomposed yields half its volume
of nitrogen.   There are, therefore, only half as
many nitrogen molecules as there were molecules
of ammonia.   It follows that each molecule of
nitrogen gas has received a contribution of nitrogen
from *two* molecules of ammonia.   If the molecule
of ammonia contain only one chemical atom of
nitrogen, then the molecule of nitrogen gas contains two chemical atoms, and no more.   It would
lead us too far to discuss the evidence, which
appears to show that the molecule of ammonia
contains only one chemical atom of nitrogen.

According to the view at first adopted by
chemists, that the chemical atom of an elementary
gas was identical with the molecule, it would follow
that *no compound* (such as ammonia) *could yield
less than its own volume of any gaseous constituent*
(such as nitrogen).   Experiments such as that in
Exercise XV. show that the ultimate particles of
an elementary substance, even in the attenuated
form of a gas, may consist, like those of compound
bodies, of two or more chemical atoms united
together.   The properties of an elementary substance constituted in this manner depend upon the
properties of a *group* of chemical atoms.   The
atoms of nitrogen appear to be firmly united in
the molecule of nitrogen gas, and this two-atom
group shows little chemical activity ; it does not
readily take part in chemical reactions.   Nitrogen

atoms in many chemical compounds, on the other
hand, readily take part in reactions. Under some
conditions, therefore, the atom of nitrogen is
chemically active, although, when it is joined up to
a second atom of its own kind in nitrogen gas, it
is chemically inert. The learner must be upon
his guard against the confusion which sometimes
arises from the promiscuous use of the phrase 'the
properties of nitrogen,' or 'the properties of
oxygen,' to denote *either* the properties of nitrogen
gas (or oxygen gas), *or* the properties of the
chemical atom of the element.

*Upon the relative weights of the chemical atoms ;
upon the choice of a unit to which the weights of the
atoms are referred ; and upon Prout's hypothesis.*—
The study of the densities of gases and of the
volumetric proportions in which gases react, showed
that chemical elements do not always unite atom
for atom, but that the atom of one element may be
equivalent to more than one atom of another
element. Now in the case of hydrogen, we do not
know of an instance in which the atom is equivalent to more than one atom of another element.
Further, the *equivalent weight* of hydrogen is less
than that of any other known element. It appeared, therefore, to be logical and convenient to
choose the equivalent weight of hydrogen as unity
(or 'weight 1') in tables of equivalent weights,
and also to choose the weight of the atom of
hydrogen as unity in the table of 'atomic weights,'
i.e. the table expressing the relative weights of the

chemical atoms. Thus, if the atoms of chlorine, oxygen, nitrogen, and carbon combine respectively with one, two, three, and four atoms of hydrogen, we have the following numbers :—

	Equivalent weights (as given by standard determinations)	Multiplied by	Atomic weights
Hydrogen	1	1	1
Chlorine	35·37	1	35·37
Oxygen	7·98	2	15·96
Nitrogen	4·67	3	14·01
Carbon	2·992	4	11·97

It would be beyond the scope of this book to discuss in detail the considerations which assist in the determination of the *multiple* of the equivalent weight which gives the atomic weight. The most important consideration is the weight of the element in the molecules of its gaseous compounds. The determination of the density of a compound gas gives the weight of its molecule, and analysis gives the per cent. of this weight which is due to the element in question. The smallest weight of the whole of an element present in the molecule of any of its compounds, or, the least difference between the weights of the element present in the molecules of its compounds gives a probable value for the atomic weight.

Hydrogen is the lightest gas known, and its molecule contains two atoms. It appears, therefore, to be logical to reckon the density of hydrogen as the unit in the table of the relative density of gases. If a gas is stated to have a density of 20, we mean that, bulk for bulk, it is twenty times as heavy as hydrogen gas, and that its molecule is twenty times as heavy as the molecule of hydrogen, and forty times as heavy as the atom of hydrogen. The molecular weight of such gas is said to be 40, i.e. *the weight of the* ATOM *of hydrogen is taken as the unit in the table of* MOLECULAR *weights* as well as in that of atomic weights.

There are, however, practical objections to the logical system of reckoning the atom of hydrogen as the unit weight. Few of the equivalent (and hence of the atomic) weights are directly determined relatively to that of hydrogen, whereas a great number are determined relatively to that of oxygen, as, for instance, by converting a metal into its oxide or reducing an oxide to metal. The result of each such experiment has to be combined with the determination of the combining proportion of hydrogen and oxygen in order to calculate the equivalent, or the atomic, weight of the element in question in terms of that of hydrogen. Thus any alteration in the received experimental numbers for the composition by weight of water alters a very large proportion of the numbers expressing atomic weights. This is highly inconvenient, as the combining proportion of oxygen and hydrogen is difficult to determine with great precision, and as successive experiments are constantly giving slightly different numbers.

The atomic weight of oxygen is *about* 16

times that of hydrogen. (Lothar Meyer and Scubert's tables give 15·96, Clarke's tables give 15·963 as the most likely value). Probably the best plan to adopt (*vide* Ostwald's 'Outlines of General Chemistry,' pp. 14–15) is to take the atomic weight of oxygen as the standard of reference, giving to it the number, not of unity, but 16. New determinations of the composition by weight of water will then only affect the precise number for hydrogen, which at present stands on the above system at 1·0032, leaving the other atomic weights practically unaffected. When this plan is adopted it is usual to state that the atomic weights are given in terms of

$$O = 16$$

The learner must bear in mind that this is simply a mode of statement adopted for convenience ; it does *not* mean that approximate numbers are employed, as when we say that the weight of the oxygen atom is about sixteen times the weight of the hydrogen atom. Further, it does *not* mean that any assumption is made that the weight of the oxygen atom is really an exact multiple of the weight of the hydrogen atom.

Prout's hypothesis.—Not long after the publication of Dalton's atomic theory many chemists inclined to the view that there existed a relation, or law, among the atomic weights similar in form to Dalton's law of multiple proportions. This idea, which has appeared in several modifications, is best known as Prout's hypothesis. The simplest and most important form of the hypothesis supposes that the weight of all the atoms are whole multiples of the weight of the atom of hydrogen.

This simple relation has not been proved.

The examination of a table of atomic weights shows, however, that there are many more than half of the elements whose atomic weights are within $\pm \frac{1}{4}$ of whole multiples of the atomic weight of hydrogen. There is a tendency to approximate towards whole numbers, but this does not show that there is any *simple* mathematical law such as regulates the combining proportions of the elements.

There has been a prepossession in favour of Prout's hypothesis due to desire for simplification. It has been thought that if the atomic weights could be shown to be integral multiples of that of hydrogen, then all elements might be regarded as condensed forms of a single stuff, which in its lightest form is hydrogen.

If, however, it should be shown that there is no such simple relation between the atomic weights, there would be nothing in the discovery to negative the existence of a single primary stuff or matter, for we have no experience of the formation of our chemical elements from any simpler materials, and we have no knowledge whether our law of conservation of mass would hold in such a process.

.

CHAPTER VIII. (*Supplementary*)

EXERCISES SUPPLEMENTARY TO THE COURSE
ILLUSTRATING THE SCOPE OF THE TERM
EQUIVALENCE IN CHEMISTRY

WHEN time and opportunity serve, it is well for
the learner to carry out a set of experiments which
will illustrate the scope of the term *equivalence* in
chemistry.

The set of experiments shown in the subjoined
diagram will serve the purpose (see table of contents at the beginning of the book). The learner
may with advantage practise himself in setting
out such diagrams in illustration of the connection
between different reactions with which he becomes acquainted in the course of his studies.

The numbers printed on the subjoined diagram
are the 'round numbers' nearest to the standard
numbers.

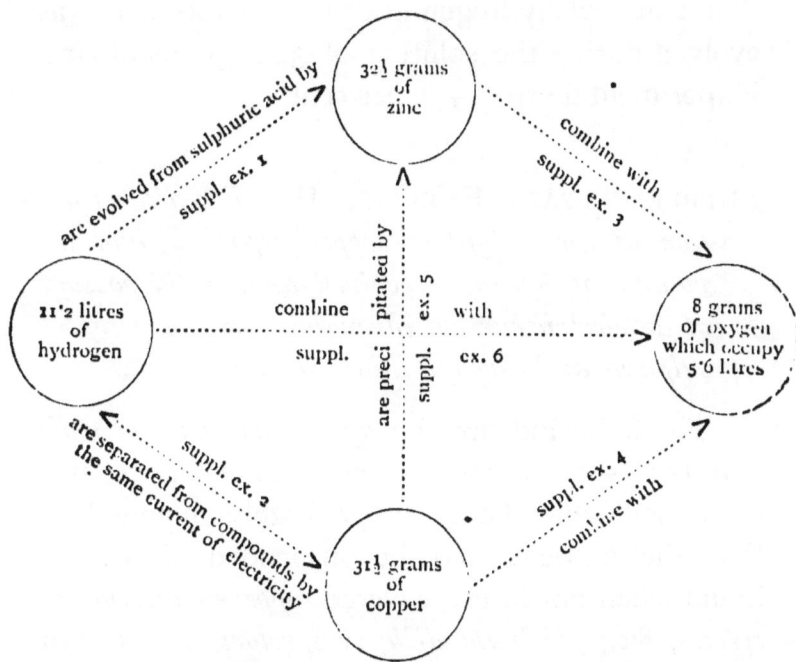

SUPPLEMENTARY EXERCISE I.—*The determination of the volume of hydrogen evolved during the solution of a known weight of zinc in dilute sulphuric acid.* (See Exercise IV. (A), Chapter III.)

The method of performing this experiment has already been described. For the purpose of the comparison with oxygen in Supplementary Exercise VI., the volume of the gas obtained should be reduced to what it would occupy at a temperature of 0° C. and at a pressure of 760 m.m. of mercury, according to the rules given in works upon physics. *Example of results obtained.*—A pupil found that 1 litre of hydrogen at 0° C. and 760 m.m. was evolved during the solution of 32·8 grams of zinc. Experimental error + 1 per cent.

SUPPLEMENTARY EXERCISE II.—*The determination of the weight of copper deposited, and the volume of hydrogen evolved, during the passage of an electric current through a solution of copper sulphate and through dilute sulphuric acid.*

The following are the *apparatus and materials required:*—An *electric battery*, three or four bichromate cells may be used, preferably mounted so that the carbons may be suspended above the liquid when not in use, covered *copper-wire, binding screws*, &c., *platinum dish*, two *retort stands* with rings and clamps, two glass *funnels*, one of them having the neck cut short and stoppered with a paraffined cork through which pass platinum electrodes, *gas-measuring tube* as in the last exercise, dilute *sulphuric acid* and pure *copper sulphate, alcohol, ether.* It is also very desirable to have a box of *resistance coils*, and a *key* for starting and stopping the current.

Mode of conducting the experiment.—The gas-measuring tube, filled with dilute sulphuric acid, is inverted in the stoppered funnel, which is supported on the ring of a retort stand, and the end of the tube is brought over one of the platinum electrodes, in which position it is clamped. The *other* platinum electrode is connected with the carbon end of the

battery. Care must be taken that the measuring
tube is placed so that no bubbles of oxygen can
mix with the hydrogen which will be evolved from
the electrode covered by the measuring tube. The
weighed platinum dish containing a solution of
pure copper sulphate is supported upon the ring of
the second retort stand. In the solution of copper
sulphate there dips a copper electrode which is
connected with the electrode under the gas-
measuring tube. When the current is passing
oxygen will be evolved from the solution of copper
sulphate, and in order to avoid spirting, it is well
to support an inverted glass funnel above the
platinum dish, the platinum wire attached to the
copper electrode passing down the neck of the
funnel. When all is ready, the platinum dish is
connected with the zinc end of the battery, the
current passes, copper deposits on the platinum
dish, and hydrogen gas collects in the measuring
tube. It is advisable to adjust the resistance of
the circuit so that the evolution of gas is slow ;
good results have been obtained with a current
giving 2 cb.c. of hydrogen per minute. The current
should be allowed to pass until not less than
60 cb.c. of hydrogen have been collected. The
gas is measured in the usual way, the temperature
and the height of the barometer being noted.
The platinum dish with its deposit of copper is
weighed. Precautions must be taken to prevent
the film of copper from oxidising, as readily
happens if it be heated. A good way of washing

H

and drying the copper film is to pour off the solution of copper sulphate, pour into the dish cold distilled water and decant this off once or twice, then wash out the water with alcohol, and then wash out the alcohol with a little ether. The ether is readily removed by blowing air from a bellows into the dish. The dish can then be weighed.

Examples of results obtained.—Pupils obtained results for the weight of copper corresponding to 11·2 litres of hydrogen, which differed from the standard numbers by − ·6 per cent., − 1 per cent., + 5 per cent., and − ·35 per cent.

SUPPLEMENTARY EXERCISE III.—*The determination of the proportion by weight in which zinc and oxygen combine.*

A weighed quantity (nearly 1 gram) of pure zinc may be dissolved in dilute sulphuric acid, excess carbonate of ammonia added, and the whole evaporated down to dryness. The substance may then be ignited strongly in a loosely covered porcelain crucible till the weight is constant, when the carbonate has been converted into oxide.

Another method of converting zinc into the oxide is to dissolve the metal in nitric acid, evaporate to dryness, and heat strongly till the weight is constant.

SUPPLEMENTARY EXERCISE IV.—*The determination of the proportion by weight in which copper and oxygen combine.*

This may be carried out by converting pure electrolytic copper to oxide as follows. Dissolve the copper in nitric acid, add solution of caustic potash to the hot solution of the nitrate. For details of the method of collecting and weighing the oxide of copper, consult Thorpe's 'Quantitative Analysis,' under the head of ' Copper Sulphate,' or any similar work on analysis.

Example of results obtained.—A pupil found that ·5025 grams copper combined with ·1272 grams oxygen. Therefore 31·59 parts of copper combine with 7·996 parts of oxygen. The standard number is 7·98, therefore the experimental error is + ·2 per cent.

SUPPLEMENTARY EXERCISE V.—*The determination of the proportion by weight between the zinc dissolved and the copper deposited when pure metallic zinc is placed in a solution of a copper salt, the copper salt being in excess.*

The experiment is done in a platinum dish. The copper deposited is dried and weighed as in Supplementary Exercise II.

SUPPLEMENTARY EXERCISE VI.—*The determination of the proportion in which hydrogen and oxygen combine.*

The direct determination of the combining weights of hydrogen and oxygen by Dumas's method (reduction of copper oxide in hydrogen and weighing the water formed) is not easy to conduct. The volumetric proportion in which hydrogen and oxygen combine may be more readily determined if suitable apparatus be at hand. This determination is constantly made in the ordinary operations of gas analysis, and shows that 5·6 litres of oxygen combine with 11·2 litres of hydrogen. If this result be obtained, it will remain in order to complete the set of experiments shown in the diagram, to show that 5·6 litres of oxygen gas weigh 8 grams, the weight which was found to combine with $31\frac{1}{2}$ grams of copper and with $32\frac{1}{2}$ grams of zinc. For this purpose the pupil may repeat the well-known exercise of heating potassium chlorate, determining the volume of water expelled from an aspirator, and noting the loss of weight suffered by the potassium chlorate.

PRINTED BY
SPOTTISWOODE AND CO., NEW-STREET SQUARE
LONDON

www.ingramcontent.com/pod-product-compliance
Lightning Source LLC
Chambersburg PA
CBHW021944190326
41519CB00009B/1136